ISACA®
Trust in, and value from, information systems

COBIT®

5

A Business Framework for the Governance and Management of Enterprise IT

COBIT® 5

AN ISACA® FRAMEWORK

ISACA®

With 95,000 constituents in 160 countries, ISACA (*www.isaca.org*) is a leading global provider of knowledge, certifications, community, advocacy and education on information systems (IS) assurance and security, enterprise governance and management of IT, and IT-related risk and compliance. Founded in 1969, the non-profit, independent ISACA hosts international conferences, publishes the *ISACA® Journal*, and develops international IS auditing and control standards, which help its constituents ensure trust in, and value from, information systems. It also advances and attests IT skills and knowledge through the globally respected Certified Information Systems Auditor® (CISA®), Certified Information Security Manager® (CISM®), Certified in the Governance of Enterprise IT® (CGEIT®) and Certified in Risk and Information Systems Control™ (CRISC™) designations. ISACA continually updates COBIT®, which helps IT professionals and enterprise leaders fulfil their IT governance and management responsibilities, particularly in the areas of assurance, security, risk and control, and deliver value to the business.

Disclaimer

ISACA has designed this publication, COBIT® 5 (the 'Work'), primarily as an educational resource for governance of enterprise IT (GEIT), assurance, risk and security professionals. ISACA makes no claim that use of any of the Work will assure a successful outcome. The Work should not be considered inclusive of all proper information, procedures and tests or exclusive of other information, procedures and tests that are reasonably directed to obtaining the same results. In determining the propriety of any specific information, procedure or test, readers should apply their own professional judgement to the specific GEIT, assurance, risk and security circumstances presented by the particular systems or information technology environment.

Copyright

ISACA

3701 Algonquin Road, Suite 1010
Rolling Meadows, IL 60008 USA
Phone: +1.847.253.1545
Fax: +1.847.253.1443
Email: *info@isaca.org*
Web site: *www.isaca.org*

Feedback: *www.isaca.org/cobit*
Participate in the ISACA Knowledge Center: *www.isaca.org/knowledge-center*
Follow ISACA on Twitter: *https://twitter.com/ISACANews*
Join the COBIT conversation on Twitter: #COBIT
Join ISACA on LinkedIn: ISACA (Official), *http://linkd.in/ISACAOfficial*
Like ISACA on Facebook: *www.facebook.com/ISACAHQ*

COBIT® 5
ISBN 978-1-60420-237-3
Printed in the United States of America
5

ACKNOWLEDGEMENTS

ISACA wishes to recognise:

COBIT 5 Task Force (2009–2011)
John W. Lainhart, IV, CISA, CISM, CGEIT, IBM Global Business Services, USA, Co-chair
Derek J. Oliver, Ph.D., DBA, CISA, CISM, CRISC, CITP, FBCS, FISM, MInstISP,
    Ravenswood Consultants Ltd., UK, Co-chair
Pippa G. Andrews, CISA, ACA, CIA, KPMG, Australia
Elisabeth Judit Antonsson, CISM, Nordea Bank, Sweden
Steven A. Babb, CGEIT, CRISC, Betfair, UK
Steven De Haes, Ph.D., University of Antwerp Management School, Belgium
Peter Harrison, CGEIT, FCPA, IBM Australia Ltd., Australia
Jimmy Heschl, CISA, CISM, CGEIT, ITIL Expert, bwin.party digital entertainment plc, Austria
Robert D. Johnson, CISA, CISM, CGEIT, CRISC, CISSP, Bank of America, USA
Erik H.J.M. Pols, CISA, CISM, Shell International-ITCI, The Netherlands
Vernon Richard Poole, CISM, CGEIT, Sapphire, UK
Abdul Rafeq, CISA, CGEIT, CIA, FCA, A. Rafeq and Associates, India

Development Team
Floris Ampe, CISA, CGEIT, CIA, ISO 27000, PwC, Belgium
Gert du Preez, CGEIT, PwC, Canada
Stefanie Grijp, PwC, Belgium
Gary Hardy, CGEIT, IT Winners, South Africa
Bart Peeters, PwC, Belgium
Geert Poels, Ghent University, Belgium
Dirk Steuperaert, CISA, CGEIT, CRISC, IT In Balance BVBA, Belgium

Workshop Participants
Gary Baker, CGEIT, CA, Canada
Brian Barnier, CGEIT, CRISC, ValueBridge Advisors, USA
Johannes Hendrik Botha, MBCS-CITP, FSM, getITright Skills Development, South Africa
Ken Buechler, CGEIT, CRISC, PMP, Great-West Life, Canada
Don Caniglia, CISA, CISM, CGEIT, FLMI, USA
Mark Chaplin, UK
Roger Debreceny, Ph.D., CGEIT, FCPA, University of Hawaii at Manoa, USA
Mike Donahue, CISA, CISM, CGEIT, CFE, CGFM, CICA, Towson University, USA
Urs Fischer, CISA, CRISC, CPA (Swiss), Fischer IT GRC Consulting & Training, Switzerland
Bob Frelinger, CISA, CGEIT, Oracle Corporation, USA
James Golden, CISM, CGEIT, CRISC, CISSP, IBM, USA
Meenu Gupta, CISA, CISM, CBP, CIPP, CISSP, Mittal Technologies, USA
Gary Langham, CISA, CISM, CGEIT, CISSP, CPFA, Australia
Nicole Lanza, CGEIT, IBM, USA
Philip Le Grand, PRINCE2, Ideagen Plc, UK
Debra Mallette, CISA, CGEIT, CSSBB, Kaiser Permanente IT, USA
Stuart MacGregor, Real IRM Solutions (Pty) Ltd., South Africa
Christian Nissen, CISM, CGEIT, FSM, CFN People, Denmark
Jamie Pasfield, ITIL V3, MSP, PRINCE2, Pfizer, UK
Eddy J. Schuermans, CGEIT, ESRAS bvba, Belgium
Michael Semrau, RWE Germany, Germany
Max Shanahan, CISA, CGEIT, FCPA, Max Shanahan & Associates, Australia
Alan Simmonds, TOGAF9, TCSA, PreterLex, UK
Cathie Skoog, CISM, CGEIT, CRISC, IBM, USA
Dejan Slokar, CISA, CGEIT, CISSP, Deloitte & Touche LLP, Canada
Roger Southgate, CISA, CISM, UK
Nicky Tiesenga, CISA, CISM, CGEIT, CRISC, IBM, USA
Wim Van Grembergen, Ph.D., University of Antwerp Management School, Belgium
Greet Volders, CGEIT, Voquals N.V., Belgium
Christopher Wilken, CISA, CGEIT, PwC, USA
Tim M. Wright, CISA, CRISC, CBCI, GSEC, QSA, Kingston Smith Consulting LLP, UK

ACKNOWLEDGEMENTS *(CONT.)*

Expert Reviewers
Mark Adler, CISA, CISM, CGEIT, CRISC, Commercial Metals Company, USA
Wole Akpose, Ph.D., CGEIT, CISSP, Morgan State University, USA
Krzysztof Baczkiewicz, CSAM, CSOX, Eracent, Poland
Roland Bah, CISA, MTN Cameroon, Cameroon
Dave Barnett, CISSP, CSSLP, USA
Max Blecher, CGEIT, Virtual Alliance, South Africa
Ricardo Bria, CISA, CGEIT, CRISC, Meycor GRC, Argentina
Dirk Bruyndonckx, CISA, CISM, CGEIT, CRISC, MCA, KPMG Advisory, Belgium
Donna Cardall, UK
Debra Chiplin, Investors Group, Canada
Sara Cosentino, CA, Great-West Life, Canada
Kamal N. Dave, CISA, CISM, CGEIT, Hewlett Packard, USA
Philip de Picker, CISA, MCA, National Bank of Belgium, Belgium
Abe Deleon, CISA, IBM, USA
Stephen Doyle, CISA, CGEIT, Department of Human Services, Australia
Heidi L. Erchinger, CISA, CRISC, CISSP, System Security Solutions, Inc., USA
Rafael Fabius, CISA, CRISC, Uruguay
Urs Fischer, CISA, CRISC, CPA (Swiss), Fischer IT GRC Consulting & Training, Switzerland
Bob Frelinger, CISA, CGEIT, Oracle Corporation, USA
Yalcin Gerek, CISA, CGEIT, CRISC, ITIL Expert, ITIL V3 Trainer, PRINCE2, ISO/IEC 20000 Consultant, Turkey
Edson Gin, CISA, CISM, CFE, CIPP, SSCP, USA
James Golden, CISM, CGEIT, CRISC, CISSP, IBM, USA
Marcelo Hector Gonzalez, CISA, CRISC, Banco Central Republic Argentina, Argentina
Erik Guldentops, University of Antwerp Management School, Belgium
Meenu Gupta, CISA, CISM, CBP, CIPP, CISSP, Mittal Technologies, USA
Angelica Haverblad, CGEIT, CRISC, ITIL, Verizon Business, Sweden
Kim Haverblad, CISM, CRISC, PCI QSA, Verizon Business, Sweden
J. Winston Hayden, CISA, CISM, CGEIT, CRISC, South Africa
Eduardo Hernandez, ITIL V3, HEME Consultores, Mexico
Jorge Hidalgo, CISA, CISM, CGEIT, ATC, Lic. Sistemas, Argentina
Michelle Hoben, Media 24, South Africa
Linda Horosko, Great-West Life, Canada
Mike Hughes, CISA, CGEIT, CRISC, 123 Consultants, UK
Grant Irvine, Great-West Life, Canada
Monica Jain, CGEIT, CSQA, CSSBB, Southern California Edison, USA
John E. Jasinski, CISA, CGEIT, SSBB, ITIL Expert, USA
Masatoshi Kajimoto, CISA, CRISC, Japan
Joanna Karczewska, CISA, Poland
Kamal Khan, CISA, CISSP, CITP, Saudi Aramco, Saudi Arabia
Eddy Khoo S. K., Prudential Services Asia, Malaysia
Marty King, CISA, CGEIT, CPA, Blue Cross Blue Shield NC, USA
Alan S. Koch, ITIL Expert, PMP, ASK Process Inc., USA
Gary Langham, CISA, CISM, CGEIT, CISSP, CPFA, Australia
Jason D. Lannen, CISA, CISM, TurnKey IT Solutions, LLC, USA
Nicole Lanza, CGEIT, IBM, USA
Philip Le Grand, PRINCE2, Ideagen Plc, UK
Kenny Lee, CISA, CISM, CISSP, Bank of America, USA
Brian Lind, CISA, CISM, CRISC, Topdanmark Forsikring A/S, Denmark
Bjarne Lonberg, CISSP, ITIL, A.P. Moller - Maersk, Denmark
Stuart MacGregor, Real IRM Solutions (Pty) Ltd., South Africa
Debra Mallette, CISA, CGEIT, CSSBB, Kaiser Permanente IT, USA
Charles Mansour, CISA, Charles Mansour Audit & Risk Service, UK
Cindy Marcello, CISA, CPA, FLMI, Great-West Life & Annuity, USA
Nancy McCuaig, CISSP, Great-West Life, Canada
John A. Mitchell, Ph.D., CISA, CGEIT, CEng, CFE, CITP, FBCS, FCIIA, QiCA, LHS Business Control, UK
Makoto Miyazaki, CISA, CPA, Bank of Tokyo-Mitsubishi, UFJ Ltd., Japan

ACKNOWLEDGEMENTS *(CONT.)*

Expert Reviewers *(cont.)*
Lucio Augusto Molina Focazzio, CISA, CISM, CRISC, ITIL, Independent Consultant, Colombia
Christian Nissen, CISM, CGEIT, FSM, ITIL Expert, CFN People, Denmark
Tony Noblett, CISA, CISM, CGEIT, CISSP, USA
Ernest Pages, CISA, CGEIT, MCSE, ITIL, Sciens Consulting LLC, USA
Jamie Pasfield, ITIL V3, MSP, PRINCE2, Pfizer, UK
Tom Patterson, CISA, CGEIT, CRISC, CPA, IBM, USA
Robert Payne, CGEIT, MBL, MCSSA, PrM, Lode Star Strategy Consulting, South Africa
Andy Piper, CISA, CISM, CRISC, PRINCE2, ITIL, Barclays Bank Plc, UK
Andre Pitkowski, CGEIT, CRISC, OCTAVE, ISO27000LA, ISO31000LA, APIT Consultoria de Informatica Ltd., Brazil
Dirk Reimers, Hewlett-Packard, Germany
Steve Reznik, CISA, ADP, Inc., USA
Robert Riley, CISSP, University of Notre Dame, USA
Martin Rosenberg, Ph.D., Cloud Governance Ltd., UK
Claus Rosenquist, CISA, CISSP, Nets Holding, Denmark
Jeffrey Roth, CISA, CGEIT, CISSP, L-3 Communications, USA
Cheryl Santor, CISSP, CNA, CNE, Metropolitan Water District, USA
Eddy J. Schuermans, CGEIT, ESRAS bvba, Belgium
Michael Semrau, RWE Germany, Germany
Max Shanahan, CISA, CGEIT, FCPA, Max Shanahan & Associates, Australia
Alan Simmonds, TOGAF9, TCSA, PreterLex, UK
Dejan Slokar, CISA, CGEIT, CISSP, Deloitte & Touche LLP, Canada
Jennifer Smith, CISA, CIA, Salt River Pima Maricopa Indian Community, USA
Marcel Sorouni, CISA, CISM, CISSP, ITIL, CCNA, MCDBA, MCSE, Bupa Australia, Australia
Roger Southgate, CISA, CISM, UK
Mark Stacey, CISA, FCA, BG Group Plc, UK
Karen Stafford Gustin, MLIS, London Life Insurance Company, Canada
Delton Sylvester, Silver Star IT Governance Consulting, South Africa
Katalin Szenes, CISA, CISM, CGEIT, CISSP, University Obuda, Hungary
Halina Tabacek, CGEIT, Oracle Americas, USA
Nancy Thompson, CISA, CISM, CGEIT, IBM, USA
Kazuhiro Uehara, CISA, CGEIT, CIA, Hitachi Consulting Co., Ltd., Japan
Rob van der Burg, Microsoft, The Netherlands
Johan van Grieken, CISA, CGEIT, CRISC, Deloitte, Belgium
Flip van Schalkwyk, Centre for e-Innovation, Western Cape Government, South Africa
Jinu Varghese, CISA, CISSP, ITIL, OCA, Ernst & Young, Canada
Andre Viviers, MCSE, IT Project+, Media 24, South Africa
Greet Volders, CGEIT, Voquals N.V., Belgium
David Williams, CISA, Westpac, New Zealand
Tim M. Wright, CISA, CRISC, CBCI, GSEC, QSA, Kingston Smith Consulting LLP, UK
Amanda Xu, PMP, Southern California Edison, USA
Tichaona Zororo, CISA, CISM, CGEIT, Standard Bank, South Africa

ISACA Board of Directors
Kenneth L. Vander Wal, CISA, CPA, Ernst & Young LLP (retired), USA, International President
Christos K. Dimitriadis, Ph.D., CISA, CISM, CRISC, INTRALOT S.A., Greece, Vice President
Gregory T. Grocholski, CISA, The Dow Chemical Co., USA, Vice President
Tony Hayes, CGEIT, AFCHSE, CHE, FACS, FCPA, FIIA, Queensland Government, Australia, Vice President
Niraj Kapasi, CISA, Kapasi Bangad Tech Consulting Pvt. Ltd., India, Vice President
Jeff Spivey, CRISC, CPP, PSP, Security Risk Management, Inc., USA, Vice President
Jo Stewart-Rattray, CISA, CISM, CGEIT, CRISC, CSEPS, RSM Bird Cameron, Australia, Vice President
Emil D'Angelo, CISA, CISM, Bank of Tokyo-Mitsubishi UFJ Ltd. (retired), USA, Past International President
Lynn C. Lawton, CISA, CRISC, FBCS CITP, FCA, FIIA, KPMG Ltd., Russian Federation, Past International President
Allan Neville Boardman, CISA, CISM, CGEIT, CRISC, CA (SA), CISSP, Morgan Stanley, UK, Director
Marc Vael, Ph.D., CISA, CISM, CGEIT, CISSP, Valuendo, Belgium, Director

ACKNOWLEDGEMENTS *(CONT.)*

Knowledge Board
Marc Vael, Ph.D., CISA, CISM, CGEIT, CISSP, Valuendo, Belgium, Chairman
Michael A. Berardi Jr., CISA, CGEIT, Bank of America, USA
John Ho Chi, CISA, CISM, CRISC, CBCP, CFE, Ernst & Young LLP, Singapore
Phillip J. Lageschulte, CGEIT, CPA, KPMG LLP, USA
Jon Singleton, CISA, FCA, Auditor General of Manitoba (retired), Canada
Patrick Stachtchenko, CISA, CGEIT, Stachtchenko & Associates SAS, France

Framework Committee (2009-2012)
Patrick Stachtchenko, CISA, CGEIT, Stachtchenko & Associates SAS, France, Chairman
Georges Ataya, CISA, CISM, CGEIT, CRISC, CISSP, Solvay Brussels School of Economics and Management,
    Belgium, Past Vice President
Steven A. Babb, CGEIT, CRISC, Betfair, UK
Sushil Chatterji, CGEIT, Edutech Enterprises, Singapore
Sergio Fleginsky, CISA, Akzo Nobel, Uruguay
John W. Lainhart, IV, CISA, CISM, CGEIT, CRISC, IBM Global Business Services, USA
Mario C. Micallef, CGEIT, CPAA, FIA, Malta
Anthony P. Noble, CISA, CCP, Viacom, USA
Derek J. Oliver, Ph.D., DBA, CISA, CISM, CRISC, CITP, FBCS, FISM, MInstISP,
    Ravenswood Consultants Ltd., UK
Robert G. Parker, CISA, CA, CMC, FCA, Deloitte & Touche LLP (retired), Canada
Rolf M. von Roessing, CISA, CISM, CGEIT, CISSP, FBCI, Forfa AG, Switzerland
Jo Stewart-Rattray, CISA, CISM, CGEIT, CRISC, CSEPS, RSM Bird Cameron, Australia
Robert E. Stroud, CGEIT, CA Inc., USA

Special Recognition
ISACA Los Angeles Chapter for its financial support

ISACA and IT Governance Institute® (ITGI®) Affiliates and Sponsors
American Institute of Certified Public Accountants
Commonwealth Association for Corporate Governance Inc.
FIDA Inform
Information Security Forum
Institute of Management Accountants Inc.
ISACA chapters
ITGI France
ITGI Japan
Norwich University
Solvay Brussels School of Economics and Management
Strategic Technology Management Institute (STMI) of the National University of Singapore
University of Antwerp Management School

Enterprise GRC Solutions Inc.
Hewlett-Packard
IBM
Symantec Corp.

TABLE OF CONTENTS

LIST OF FIGURES

Page intentionally left blank

COBIT 5: A BUSINESS FRAMEWORK FOR THE GOVERNANCE AND MANAGEMENT OF ENTERPRISE IT

The COBIT 5 publication contains the COBIT 5 framework for governing and managing enterprise IT. The publication is part of the COBIT 5 product family as shown in **figure 1**.

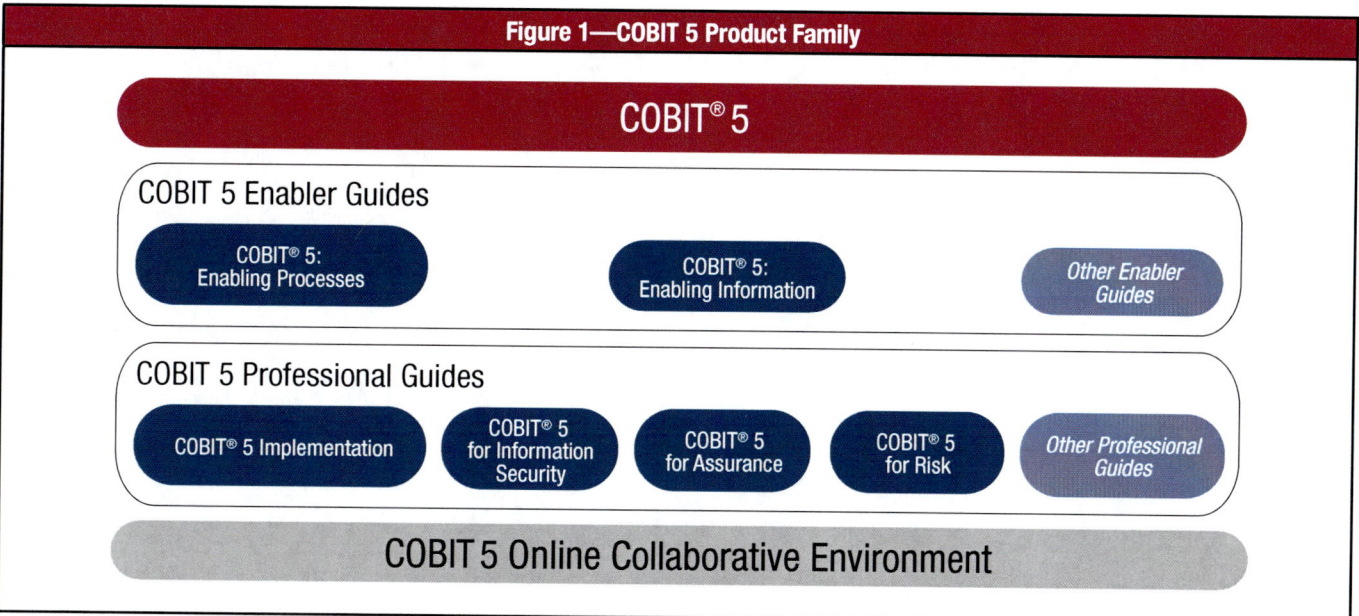

The COBIT 5 framework is built on five basic principles, which are covered in detail, and includes extensive guidance on enablers for governance and management of enterprise IT.

The COBIT 5 product family includes the following products:
• COBIT 5 (the framework)
• COBIT 5 enabler guides, in which governance and management enablers are discussed in detail. These include:
  – *COBIT 5: Enabling Processes*
  – COBIT 5: Enabling Information (in development)
  – Other enabler guides (check *www.isaca.org/cobit*)
• COBIT 5 professional guides, which include:
  – *COBIT 5 Implementation*
  – COBIT 5 for Information Security (in development)
  – COBIT 5 for Assurance (in development)
  – COBIT 5 for Risk (in development)
  – Other professional guides (check *www.isaca.org/cobit*)
• A collaborative online environment, which will be available to support the use of COBIT 5

Page intentionally left blank

EXECUTIVE SUMMARY

Information is a key resource for all enterprises, and from the time that information is created to the moment that it is destroyed, technology plays a significant role. Information technology is increasingly advanced and has become pervasive in enterprises and in social, public and business environments.

As a result, today, more than ever, enterprises and their executives strive to:
• Maintain high-quality information to support business decisions.
• Generate business value from IT-enabled investments, i.e., achieve strategic goals and realise business benefits through effective and innovative use of IT.
• Achieve operational excellence through the reliable and efficient application of technology.
• Maintain IT-related risk at an acceptable level.
• Optimise the cost of IT services and technology.
• Comply with ever-increasing relevant laws, regulations, contractual agreements and policies.

Over the past decade, the term 'governance' has moved to the forefront of business thinking in response to examples demonstrating the importance of good governance and, on the other end of the scale, global business mishaps.

Successful enterprises have recognised that the board and executives need to embrace IT like any other significant part of doing business. Boards and management—both in the business and IT functions—must collaborate and work together, so that IT is included within the governance and management approach. In addition, legislation is increasingly being passed and regulations implemented to address this need.

COBIT 5 provides a comprehensive framework that assists enterprises in achieving their objectives for the governance and management of enterprise IT. Simply stated, it helps enterprises create optimal value from IT by maintaining a balance between realising benefits and optimising risk levels and resource use. COBIT 5 enables IT to be governed and managed in a holistic manner for the entire enterprise, taking in the full end-to-end business and IT functional areas of responsibility, considering the IT-related interests of internal and external stakeholders. COBIT 5 is generic and useful for enterprises of all sizes, whether commercial, not-for-profit or in the public sector.

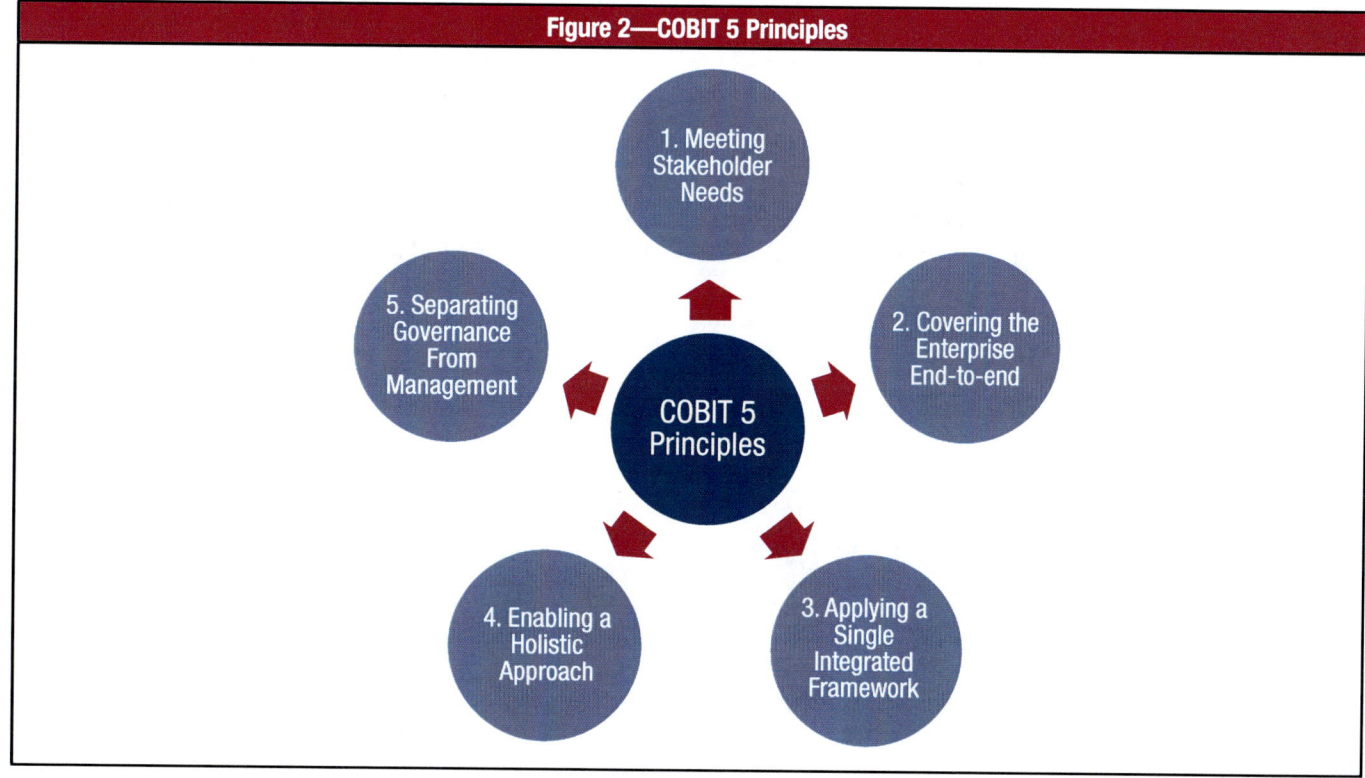

Figure 2—COBIT 5 Principles

COBIT 5 is based on five key principles (shown in **figure 2**) for governance and management of enterprise IT:
- **Principle 1: Meeting Stakeholder Needs**—Enterprises exist to create value for their stakeholders by maintaining a balance between the realisation of benefits and the optimisation of risk and use of resources. COBIT 5 provides all of the required processes and other enablers to support business value creation through the use of IT. Because every enterprise has different objectives, an enterprise can customise COBIT 5 to suit its own context through the goals cascade, translating high-level enterprise goals into manageable, specific, IT-related goals and mapping these to specific processes and practices.
- **Principle 2: Covering the Enterprise End-to-end**—COBIT 5 integrates governance of enterprise IT into enterprise governance:
  – It covers all functions and processes within the enterprise; COBIT 5 does not focus only on the 'IT function', but treats information and related technologies as assets that need to be dealt with just like any other asset by everyone in the enterprise.
  – It considers all IT-related governance and management enablers to be enterprisewide and end-to-end, i.e., inclusive of everything and everyone—internal and external—that is relevant to governance and management of enterprise information and related IT.
- **Principle 3: Applying a Single, Integrated Framework**—There are many IT-related standards and best practices, each providing guidance on a subset of IT activities. COBIT 5 aligns with other relevant standards and frameworks at a high level, and thus can serve as the overarching framework for governance and management of enterprise IT.
- **Principle 4: Enabling a Holistic Approach**—Efficient and effective governance and management of enterprise IT require a holistic approach, taking into account several interacting components. COBIT 5 defines a set of enablers to support the implementation of a comprehensive governance and management system for enterprise IT. Enablers are broadly defined as anything that can help to achieve the objectives of the enterprise. The COBIT 5 framework defines seven categories of enablers:
  – Principles, Policies and Frameworks
  – Processes
  – Organisational Structures
  – Culture, Ethics and Behaviour
  – Information
  – Services, Infrastructure and Applications
  – People, Skills and Competencies
- **Principle 5: Separating Governance From Management**—The COBIT 5 framework makes a clear distinction between governance and management. These two disciplines encompass different types of activities, require different organisational structures and serve different purposes. COBIT 5's view on this key distinction between governance and management is:
  – Governance

> **Governance ensures that stakeholder needs, conditions and options are evaluated to determine balanced, agreed-on enterprise objectives to be achieved; setting direction through prioritisation and decision making; and monitoring performance and compliance against agreed-on direction and objectives.**

In most enterprises, overall governance is the responsibility of the board of directors under the leadership of the chairperson. Specific governance responsibilities may be delegated to special organisational structures at an appropriate level, particularly in larger, complex enterprises.

  – Management

> **Management plans, builds, runs and monitors activities in alignment with the direction set by the governance body to achieve the enterprise objectives.**

In most enterprises, management is the responsibility of the executive management under the leadership of the chief executive officer (CEO).

Together, these five principles enable the enterprise to build an effective governance and management framework that optimises information and technology investment and use for the benefit of stakeholders.

CHAPTER 1
OVERVIEW OF COBIT 5

COBIT 5 provides the next generation of ISACA's guidance on the enterprise governance and management of IT. It builds on more than 15 years of practical usage and application of COBIT by many enterprises and users from business, IT, risk, security and assurance communities. The major drivers for the development of COBIT 5 include the need to:
• Provide more stakeholders a say in determining what they expect from information and related technology (what benefits at what acceptable level of risk and at what costs) and what their priorities are in ensuring that expected value is actually being delivered. Some will want short-term returns and others long-term sustainability. Some will be ready to take a high risk that others will not. These divergent and sometimes conflicting expectations need to be dealt with effectively. Furthermore, not only do these stakeholders want to be more involved, but they want more transparency regarding how this will happen and the actual results achieved.
• Address the increasing dependency of enterprise success on external business and IT parties such as outsourcers, suppliers, consultants, clients, cloud and other service providers, and on a diverse set of internal means and mechanisms to deliver the expected value
• Deal with the amount of information, which has increased significantly. How do enterprises select the relevant and credible information that will lead to effective and efficient business decisions? Information also needs to be managed effectively and an effective information model can assist.
• Deal with much more pervasive IT; it is more and more an integral part of the business. Often, it is no longer satisfactory to have IT separate even if it is aligned to the business. It needs to be an integral part of the business projects, organisational structures, risk management, policies, skills, processes, etc. The roles of the chief information officer (CIO) and the IT function are evolving. More and more people within the business functions have IT skills and are, or will be, involved in IT decisions and IT operations. IT and business will need to be better integrated.
• Provide further guidance in the area of innovation and emerging technologies; this is about creativity, inventiveness, developing new products, making the existing products more compelling to customers and reaching new types of customers. Innovation also implies streamlining product development, manufacturing and supply chain processes to deliver products to market with increasing levels of efficiency, speed and quality.
• Cover the full end-to-end business and IT functional responsibilities, and cover all aspects that lead to effective governance and management of enterprise IT, such as organisational structures, policies and culture, over and above processes
• Get better control over increasing user-initiated and user-controlled IT solutions
• Achieve enterprise:
  – Value creation through effective and innovative use of enterprise IT
  – Business user satisfaction with IT engagement and services
  – Compliance with relevant laws, regulations, contractual agreements and internal policies
  – Improved relations between business needs and IT objectives
• Connect to, and, where relevant, align with, other major frameworks and standards in the marketplace, such as Information Technology Infrastructure Library (ITIL®), The Open Group Architecture Forum (TOGAF®), Project Management Body of Knowledge (PMBOK®), PRojects IN Controlled Environments 2 (PRINCE2®), Committee of Sponsoring Organizations of the Treadway Commission (COSO) and the International Organization for Standardization (ISO) standards. This will help stakeholders understand how various frameworks, good practices and standards are positioned relative to each other and how they can be used together.
• Integrate all major ISACA frameworks and guidance, with a primary focus on COBIT, Val IT and Risk IT, but also considering the Business Model for Information Security (BMIS), the IT Assurance Framework (ITAF), the publication titled *Board Briefing on IT Governance*, and the Taking Governance Forward (TGF) resource, such that COBIT 5 covers the complete enterprise and provides a basis to integrate other frameworks, standards and practices as one single framework

Different products and other guidance covering the diverse needs of various stakeholders will be built from the main COBIT 5 knowledge base. This will happen over time, making the COBIT 5 product architecture a living document. The latest COBIT 5 product architecture can be found on the COBIT pages of the ISACA web site (*www.isaca.org/cobit*).

Overview of This Publication

The COBIT 5 framework contains seven more chapters:
• Chapter 2 elaborates on Principle 1, **Meeting Stakeholder Needs.** It introduces the COBIT 5 goals cascade. The enterprise goals for IT are used to formalise and structure the stakeholder needs. Enterprise goals can be linked to IT-related goals, and these IT-related goals can be achieved through the optimal use and execution of all enablers, including processes. This set of connecting goals is called the COBIT 5 goals cascade. The chapter also provides examples of typical governance and management questions that stakeholders may have about enterprise IT.
• Chapter 3 elaborates on Principle 2, **Covering the Enterprise End-to-end.** It explains how COBIT 5 integrates governance of enterprise IT into enterprise governance by covering all functions and processes within the enterprise.
• Chapter 4 elaborates on Principle 3, **Applying a Single Integrated Framework**, and describes briefly the COBIT 5 architecture that achieves the integration.
• Chapter 5 elaborates on Principle 4, **Enabling a Holistic Approach.** Governance of enterprise IT is systemic and supported by a set of enablers. In this chapter, enablers are introduced and a common way of looking at enablers is presented: the generic enabler model.
• Chapter 6 elaborates on Principle 5, **Separating Governance From Management**, and discusses the difference between management and governance, and how they interrelate. The high-level COBIT 5 process reference model is included as an example.
• Chapter 7 contains an introduction to **Implementation Guidance**. It describes how the appropriate environment can be created, the enablers required, typical pain points and trigger events for implementation, and the implementation and continual improvement life cycle. This chapter is based on the publication titled *COBIT® 5 Implementation*, where full details on how to implement governance of enterprise IT based on COBIT 5 can be found.
• Chapter 8 elaborates on **The COBIT 5 Process Capability Model** in the COBIT Assessment Programme approach (*www.isaca.org/cobit-assessment-programme*) scheme, how it differs from COBIT 4.1 process maturity assessments, and how users can migrate to the new approach.

The appendices contain reference information, mappings and more detailed information on specific subjects:
• Appendix A. **References** used during COBIT 5 development are listed.
• Appendix B. **Detailed Mapping Enterprise Goals—IT-related Goals** describes how enterprise goals typically are supported by one or more IT-related goals.
• Appendix C. **Detailed Mapping IT-related Goals—IT-related Processes** describes how COBIT processes support the achievement of IT-related goals.
• Appendix D. **Stakeholder Needs and Enterprise Goals** describes how typical stakeholder needs relate to COBIT 5 enterprise goals.
• Appendix E. **Mapping of COBIT 5 With the Most Relevant Related Standards and Frameworks**
• Appendix F. **Comparison Between the COBIT 5 Information Model and the COBIT 4.1 Information Criteria**
• Appendix G. **Detailed Description of the COBIT 5 Enablers** builds on chapter 5 and includes more details on the different enablers, including a detailed enabler model describing specific components, and is illustrated with a number of examples.
• Appendix H. **Glossary**

CHAPTER 2
PRINCIPLE 1: MEETING STAKEHOLDER NEEDS

Introduction

Enterprises exist to create value for their stakeholders. Consequently, any enterprise—commercial or not—will have value creation as a governance objective. Value creation means realising benefits at an optimal resource cost while optimising risk. (See **figure 3**.) Benefits can take many forms, e.g., financial for commercial enterprises or public service for government entities.

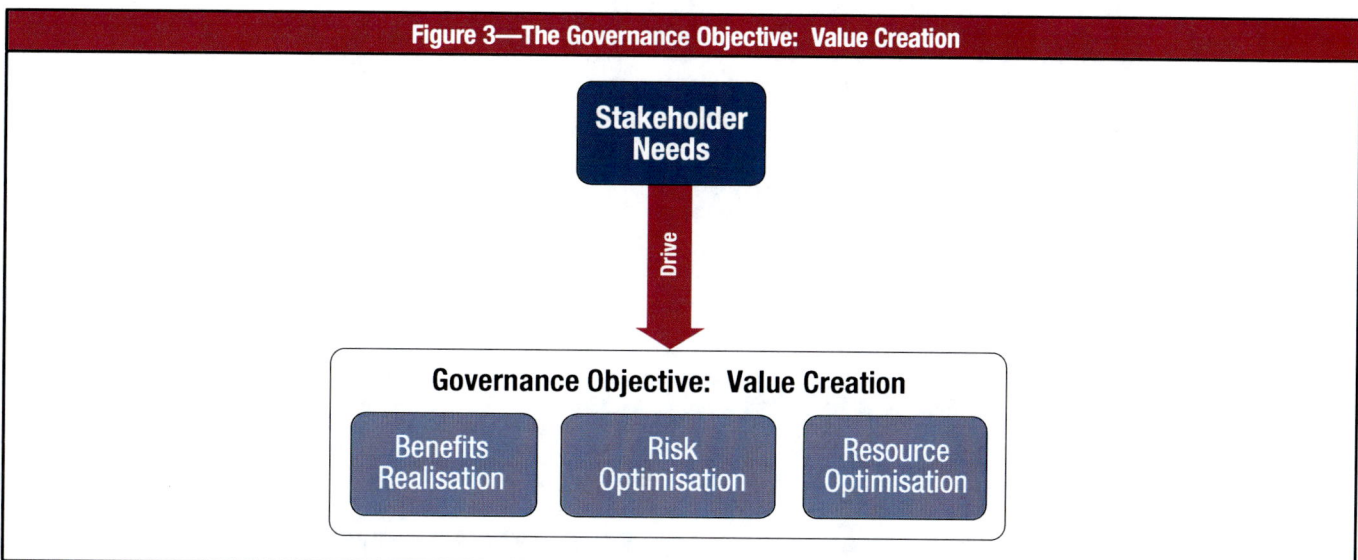

Figure 3—The Governance Objective: Value Creation

Enterprises have many stakeholders, and 'creating value' means different—and sometimes conflicting—things to each of them. Governance is about negotiating and deciding amongst different stakeholders' value interests. By consequence, the governance system should consider all stakeholders when making benefit, risk and resource assessment decisions. For each decision, the following questions can and should be asked: For whom are the benefits? Who bears the risk? What resources are required?

COBIT 5 Goals Cascade

Every enterprise operates in a different context; this context is determined by external factors (the market, the industry, geopolitics, etc.) and internal factors (the culture, organisation, risk appetite, etc.), and requires a customised governance and management system.

Stakeholder needs have to be transformed into an enterprise's actionable strategy. The COBIT 5 goals cascade is the mechanism to translate stakeholder needs into specific, actionable and customised enterprise goals, IT-related goals and enabler goals. This translation allows setting specific goals at every level and in every area of the enterprise in support of the overall goals and stakeholder requirements, and thus effectively supports alignment between enterprise needs and IT solutions and services.

The COBIT 5 goals cascade is shown in **figure 4**.

Step 1. Stakeholder Drivers Influence Stakeholder Needs
Stakeholder needs are influenced by a number of drivers, e.g., strategy changes, a changing business and regulatory environment, and new technologies.

Step 2. Stakeholder Needs Cascade to Enterprise Goals
Stakeholder needs can be related to a set of generic enterprise goals. These enterprise goals have been developed using the balanced scorecard (BSC)[1] dimensions, and they represent a list of commonly used goals that an enterprise may define for itself. Although this list is not exhaustive, most enterprise-specific goals can be mapped easily onto one or more of the generic enterprise goals. A table of stakeholder needs and enterprise goals is presented in appendix D.

[1] Kaplan, Robert S.; David P. Norton; *The Balanced Scorecard: Translating Strategy Into Action*, Harvard University Press, USA, 1996

Figure 4—COBIT 5 Goals Cascade Overview

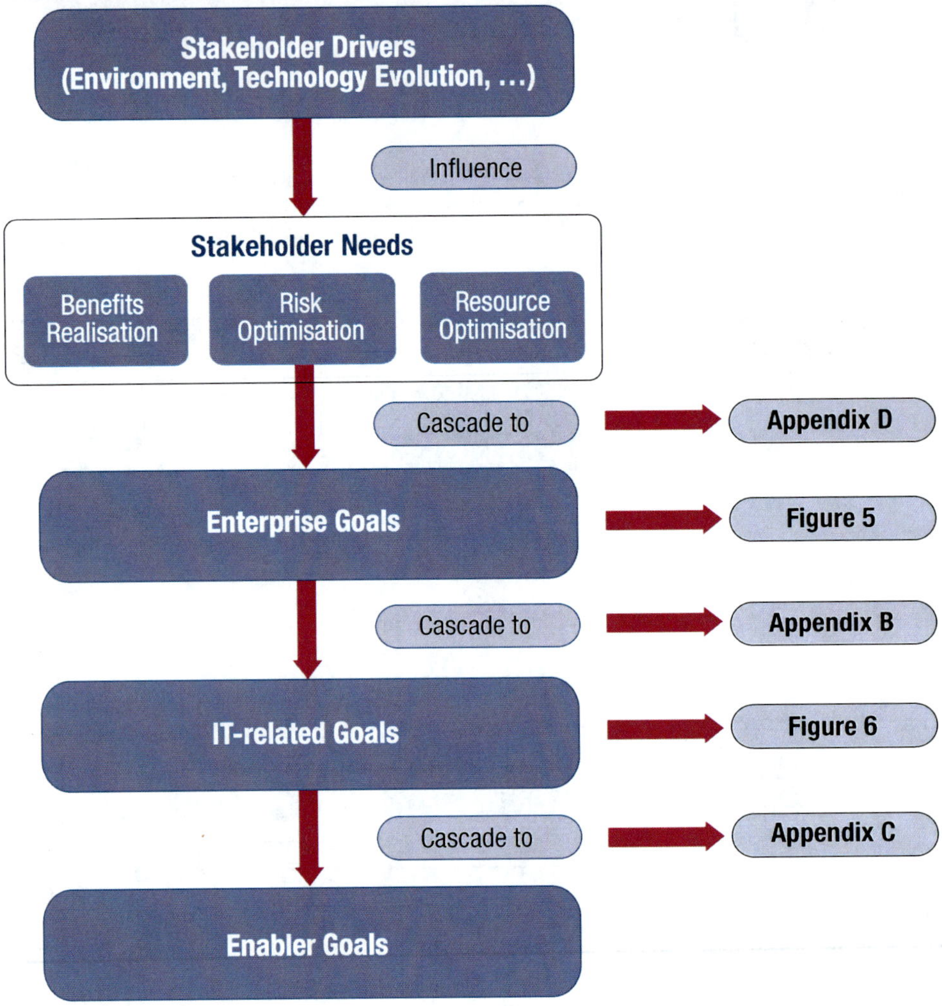

Stakeholder Drivers
(Environment, Technology Evolution, …)

Influence

Stakeholder Needs

| Benefits Realisation | Risk Optimisation | Resource Optimisation |

Cascade to → **Appendix D**

Enterprise Goals → **Figure 5**

Cascade to → **Appendix B**

IT-related Goals → **Figure 6**

Cascade to → **Appendix C**

Enabler Goals

COBIT 5 defines 17 generic goals, as shown in **figure 5**, which includes the following information:
• The BSC dimension under which the enterprise goal fits
• Enterprise goals
• The relationship to the three main governance objectives—benefits realisation, risk optimisation and resource optimisation. ('P' stands for primary relationship and 'S' for secondary relationship, i.e., a less strong relationship.)

Step 3. Enterprise Goals Cascade to IT-related Goals
Achievement of enterprise goals requires a number of IT-related outcomes,[2] which are represented by the IT-related goals. IT-related stands for information and related technology, and the IT-related goals are structured along the dimensions of the IT balanced scorecard (IT BSC). COBIT 5 defines 17 IT-related goals, listed in **figure 6**.

The mapping table between IT-related goals and enterprise goals is included in appendix B, and it shows how each enterprise goal is supported by a number of IT-related goals.

Step 4. IT-related Goals Cascade to Enabler Goals
Achieving IT-related goals requires the successful application and use of a number of enablers. The enabler concept is explained in detail in chapter 5. Enablers include processes, organisational structures and information, and for each enabler a set of specific relevant goals can be defined in support of the IT-related goals.

Processes are one of the enablers, and appendix C contains a mapping between IT-related goals and the relevant COBIT 5 processes, which then contain related process goals.

---

[2] IT-related outcomes obviously are not the only intermediate benefit required to achieve enterprise goals. All other functional areas in an organisation, such as finance and marketing, also contribute to the achievement of enterprise goals, but within the context of COBIT 5 only IT-related activities and goals are considered.

	Figure 5—COBIT 5 Enterprise Goals			
		Relation to Governance Objectives		
BSC Dimension	Enterprise Goal	Benefits Realisation	Risk Optimisation	Resource Optimisation
Financial	1. Stakeholder value of business investments	P		S
	2. Portfolio of competitive products and services	P	P	S
	3. Managed business risk (safeguarding of assets)		P	S
	4. Compliance with external laws and regulations		P	
	5. Financial transparency	P	S	S
Customer	6. Customer-oriented service culture	P		S
	7. Business service continuity and availability		P	
	8. Agile responses to a changing business environment	P		S
	9. Information-based strategic decision making	P	P	P
	10. Optimisation of service delivery costs	P		P
Internal	11. Optimisation of business process functionality	P		P
	12. Optimisation of business process costs	P		P
	13. Managed business change programmes	P	P	S
	14. Operational and staff productivity	P		P
	15. Compliance with internal policies		P	
Learning and Growth	16. Skilled and motivated people	S	P	P
	17. Product and business innovation culture	P		

	Figure 6—IT-related Goals	
IT BSC Dimension		Information and Related Technology Goal
Financial	01	Alignment of IT and business strategy
	02	IT compliance and support for business compliance with external laws and regulations
	03	Commitment of executive management for making IT-related decisions
	04	Managed IT-related business risk
	05	Realised benefits from IT-enabled investments and services portfolio
	06	Transparency of IT costs, benefits and risk
Customer	07	Delivery of IT services in line with business requirements
	08	Adequate use of applications, information and technology solutions
Internal	09	IT agility
	10	Security of information, processing infrastructure and applications
	11	Optimisation of IT assets, resources and capabilities
	12	Enablement and support of business processes by integrating applications and technology into business processes
	13	Delivery of programmes delivering benefits, on time, on budget, and meeting requirements and quality standards
	14	Availability of reliable and useful information for decision making
	15	IT compliance with internal policies
Learning and Growth	16	Competent and motivated business and IT personnel
	17	Knowledge, expertise and initiatives for business innovation

Using the COBIT 5 Goals Cascade

Benefits of the COBIT 5 Goals Cascade

The goals cascade[3] is important because it allows the definition of priorities for implementation, improvement and assurance of governance of enterprise IT based on (strategic) objectives of the enterprise and the related risk. In practice, the goals cascade:
- Defines relevant and tangible goals and objectives at various levels of responsibility
- Filters the knowledge base of COBIT 5, based on enterprise goals, to extract relevant guidance for inclusion in specific implementation, improvement or assurance projects
- Clearly identifies and communicates how (sometimes very operational) enablers are important to achieve enterprise goals

Using the COBIT 5 Goals Cascade Carefully

The goals cascade—with its mapping tables between enterprise goals and IT-related goals and between IT-related goals and COBIT 5 enablers (including processes)—does not contain the universal truth, and users should not attempt to use it in a purely mechanistic way, but rather as a guideline. There are various reasons for this, including:
- Every enterprise has different priorities in its goals, and priorities may change over time.
- The mapping tables do not distinguish between size and/or industry of the enterprise. They represent a sort of common denominator of how, in general, the different levels of goals are interrelated.
- The indicators used in the mapping use two levels of importance or relevance, suggesting that there are 'discrete' levels of relevance, whereas, in reality, the mapping will be close to a continuum of various degrees of correspondence.

Using the COBIT 5 Goals Cascade in Practice

From the previous disclaimer, it is obvious that the first step an enterprise should always apply when using the goals cascade is to customise the mapping, taking into account its specific situation. In other words, each enterprise should build its own goals cascade, compare it with COBIT and then refine it.

For example, the enterprise may wish to:
- Translate the strategic priorities into a specific 'weight' or importance for each of the enterprise goals.
- Validate the mappings of the goals cascade, taking into account its specific environment, industry, etc.

---

[3] The goals cascade is based on research performed by the University of Antwerp Management School IT Alignment and Governance Institute in Belgium.

EXAMPLE 1—GOALS CASCADE

An enterprise has defined for itself a number of strategic goals, of which improving customer satisfaction is the most important. From there, it wants to know where it needs to improve in all things related to IT.

The enterprise decides that setting customer satisfaction as a key priority is equivalent to raising the priority of the following enterprise goals (from **figure 5**):
- 6. Customer-oriented service culture
- 7. Business service continuity and availability
- 8. Agile responses to a changing business environment

The enterprise now takes the next step in the goals cascade: analysing which IT-related goals correspond to these enterprise goals. A suggested mapping between them is listed in appendix B.

From there, the following IT-related goals are suggested as most important (all 'P' relationships):
- 01 Alignment of IT and business strategy
- 04 Managed IT-related business risk
- 07 Delivery of IT services in line with business requirements
- 09 IT agility
- 10 Security of information, processing infrastructure and applications
- 14 Availability of reliable and useful information for decision making
- 17 Knowledge, expertise and initiatives for business innovation

The enterprise validates this list, and decides to retain the first four goals as a matter of priority.

In the next step in the cascade, using the enabler concept (see chapter 5), these IT-related goals drive a number of enabler goals, which include process goals. In appendix C, a mapping is suggested between IT-related goals and COBIT 5 processes. That table allows identification of the most relevant IT-related processes that support the IT-related goals, but, processes alone are not sufficient. The other enablers, such as culture, behaviour and ethics; organisational structures; or skills and expertise are equally important and require a set of clear goals.

When this exercise is completed, the enterprise has a set of consistent goals for all enablers that allow it to reach the stated strategic objectives and a set of associated metrics to measure performance.

EXAMPLE 2—STAKEHOLDER NEEDS: SUSTAINABILITY

After a stakeholder needs analysis is conducted, an enterprise decides that sustainability is a strategic priority. For it, sustainability includes not only environmental aspects, but all things that contribute to the long-term success of the enterprise.

Based on the results of the stakeholder needs analysis, the enterprise decides to focus on the following five objectives, with some further specification of the goals added:
1. Stakeholder value of business investments, especially for the stakeholders society
4. Compliance with external laws and regulations, focussing on environmental laws and laws dealing with labour regulations in outsourcing arrangements
8. Agile response to a changing business environment
16. Skilled and motivated people, recognising that the success of the enterprise depends on its people
17. Product and business innovation culture, focussing on longer-term innovations

Based on these priorities, the goals cascade can be applied as explained in the text.

Governance and Management Questions on IT

The fulfilment of stakeholder needs in any enterprise will—given the high dependency on IT—raise a number of questions on the governance and management of enterprise IT (**figure 7**).

Figure 7—Governance and Management Questions on IT	
Internal Stakeholders	**Internal Stakeholder Questions**
• Board • Chief executive officer (CEO) • Chief financial officer (CFO) • Chief information officer (CIO) • Chief risk officer (CRO) • Business executives • Business process owners • Business managers • Risk managers • Security managers • Service managers • Human resource (HR) managers • Internal audit • Privacy officers • IT users • IT managers • Etc.	• How do I get value from the use of IT? Are end users satisfied with the quality of the IT service? • How do I manage performance of IT? • How can I best exploit new technology for new strategic opportunities? • How do I best build and structure my IT department? • How dependent am I on external providers? How well are IT outsourcing agreements being managed? How do I obtain assurance over external providers? • What are the (control) requirements for information? • Did I address all IT-related risk? • Am I running an efficient and resilient IT operation? • How do I control the cost of IT? How do I use IT resources in the most effective and efficient manner? What are the most effective and efficient sourcing options? • Do I have enough people for IT? How do I develop and maintain their skills, and how do I manage their performance? • How do I get assurance over IT? • Is the information I am processing well secured? • How do I improve business agility through a more flexible IT environment? • Do IT projects fail to deliver what they promised—and if so, why? Is IT standing in the way of executing the business strategy? • How critical is IT to sustaining the enterprise? What do I do if IT is not available? • What critical business processes are dependent on IT, and what are the requirements of business processes? • What has been the average overrun of the IT operational budgets? How often and how much do IT projects go over budget? • How much of the IT effort goes to fighting fires rather than to enabling business improvements? • Are sufficient IT resources and infrastructure available to meet required enterprise strategic objectives? • How long does it take to make major IT decisions? • Are the total IT effort and investments transparent? • Does IT support the enterprise in complying with regulations and service levels? How do I know whether I am compliant with all applicable regulations?
External Stakeholders	**External Stakeholder Questions**
• Business partners • Suppliers • Shareholders • Regulators/government • External users • Customers • Standardisation organisations • External auditors • Consultants • Etc.	• How do I know my business partner's operations are secure and reliable? • How do I know the enterprise is compliant with applicable rules and regulations? • How do I know the enterprise is maintaining an effective system of internal control? • Do business partners have the information chain between them under control?

How to Find an Answer to These Questions

All questions mentioned in **figure 7** can be related to the enterprise goals, and serve as input to the goals cascade, upon which they can be addressed effectively. Appendix D contains an example mapping between the internal stakeholder questions mentioned in **figure 7** and enterprise goals.

CHAPTER 3
PRINCIPLE 2: COVERING THE ENTERPRISE END-TO-END

COBIT 5 addresses the governance and management of information and related technology from an enterprisewide, end-to-end perspective. This means that COBIT 5:
- Integrates governance of enterprise IT into enterprise governance. That is, the governance system for enterprise IT proposed by COBIT 5 integrates seamlessly in any governance system. COBIT 5 aligns with the latest views on governance.
- Covers all functions and processes required to govern and manage enterprise information and related technologies wherever that information may be processed. Given this extended enterprise scope, COBIT 5 addresses all the relevant internal and external IT services, as well as internal and external business processes.

COBIT 5 provides a holistic and systemic view on governance and management of enterprise IT (see principle 4), based on a number of enablers. The enablers are enterprisewide and end-to-end, i.e., inclusive of everything and everyone, internal and external, that are relevant to governance and management of enterprise information and related IT, including the activities and responsibilities of both the IT functions and non-IT business functions.

Information is one of the COBIT enabler categories. The model by which COBIT 5 defines enablers allows every stakeholder to define extensive and complete requirements for information and the information processing life cycle, thus connecting the business and its need for adequate information and the IT function, and supporting the business and context focus.

Governance Approach

The end-to-end governance approach that is at the foundation of COBIT 5 is depicted in **figure 8**, showing the key components of a governance system.[4]

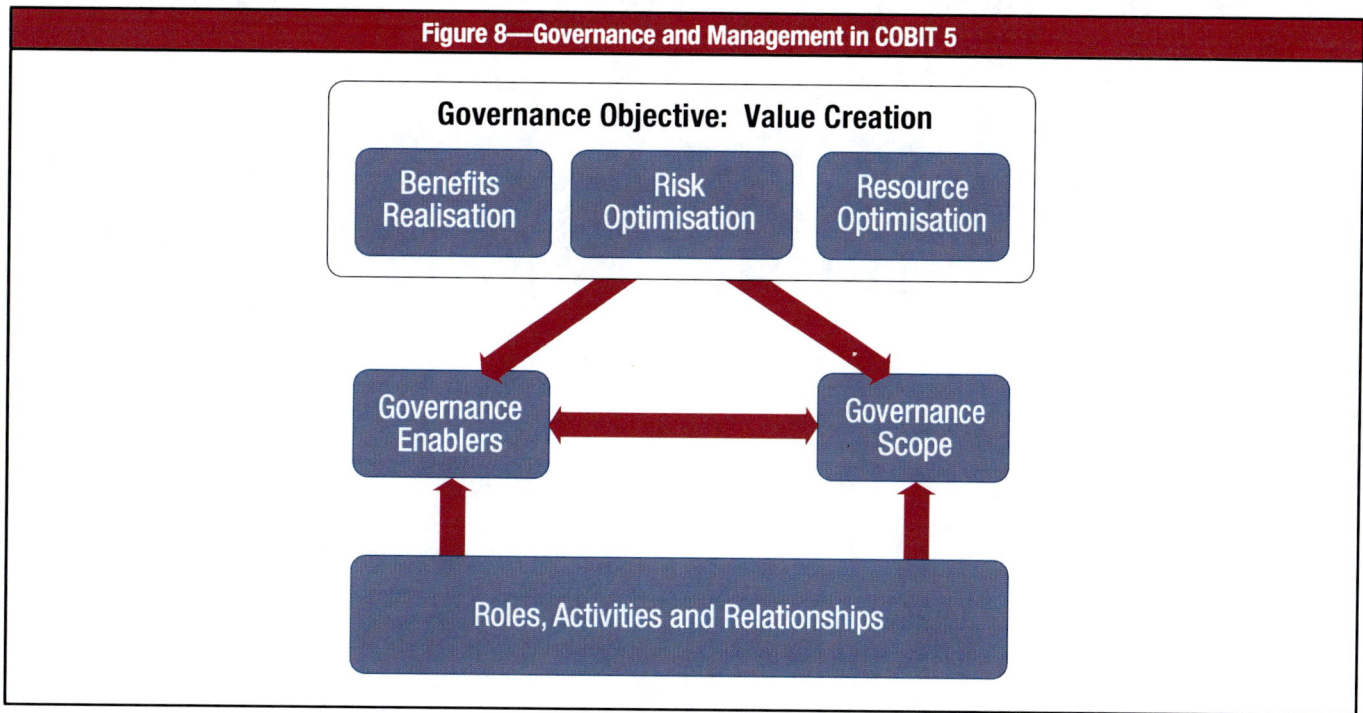

Figure 8—Governance and Management in COBIT 5

Governance Objective: Value Creation

Benefits Realisation

Risk Optimisation

Resource Optimisation

Governance Enablers

Governance Scope

Roles, Activities and Relationships

---

[4] This governance system is an illustration of ISACA's Taking Governance Forward (TGF) initiative; more information on TGF can be found at *www.takinggovernanceforward.org.*

In addition to the governance objective, the other main elements of the governance approach include enablers; scope; and roles, activities, and relationships.

Governance Enablers

Governance enablers are the organisational resources for governance, such as frameworks, principles, structures, processes and practices, through or towards which action is directed and objectives can be attained. Enablers also include the enterprise's resources—e.g., service capabilities (IT infrastructure, applications, etc.), people and information. A lack of resources or enablers may affect the ability of the enterprise to create value.

Given the importance of governance enablers, COBIT 5 includes a single way of looking at and dealing with enablers (see chapter 5).

Governance Scope

Governance can be applied to the entire enterprise, an entity, a tangible or intangible asset, etc. That is, it is possible to define different views of the enterprise to which governance is applied, and it is essential to define this scope of the governance system well. The scope of COBIT 5 is the enterprise—but in essence COBIT 5 can deal with any of the different views.

Roles, Activities and Relationships

A last element is governance roles, activities and relationships. It defines who is involved in governance, how they are involved, what they do and how they interact, within the scope of any governance system. In COBIT 5, clear differentiation is made between governance and management activities in the governance and management domains, as well as the interfacing between them and the role players that are involved. **Figure 9** details the lower part of **figure 8**, listing the interactions between the different roles.

For more information on this generic view on governance please see Taking Governance Forward at *www.takinggovernanceforward.org.*

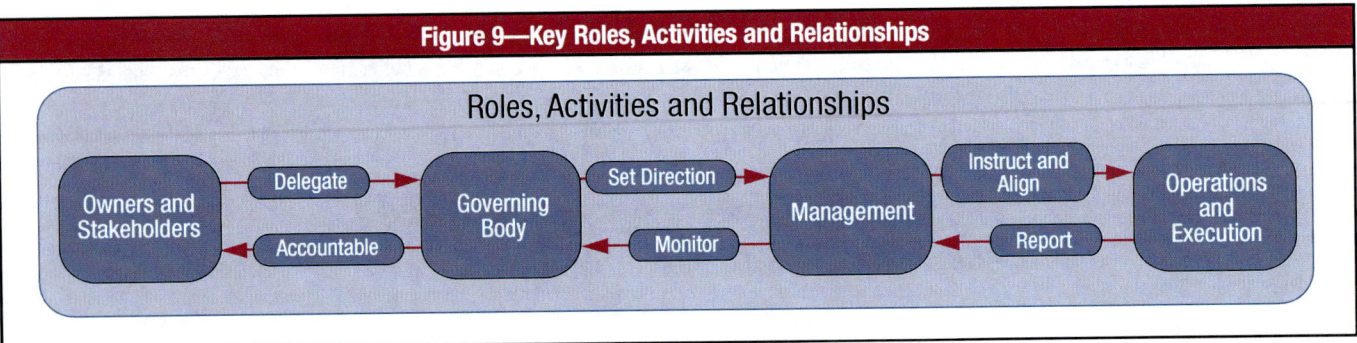

Figure 9—Key Roles, Activities and Relationships

CHAPTER 4
PRINCIPLE 3: APPLYING A SINGLE INTEGRATED FRAMEWORK

COBIT 5 is a single and integrated framework because:
- It aligns with other latest relevant standards and frameworks, and thus allows the enterprise to use COBIT 5 as the overarching governance and management framework integrator.
- It is complete in enterprise coverage, providing a basis to integrate effectively other frameworks, standards and practices used. A single overarching framework serves as a consistent and integrated source of guidance in a non-technical, technology-agnostic common language.
- It provides a simple architecture for structuring guidance materials and producing a consistent product set.
- It integrates all knowledge previously dispersed over different ISACA frameworks. ISACA has researched the key area of enterprise governance for many years and has developed frameworks such as COBIT, Val IT, Risk IT, BMIS, the publication *Board Briefing on IT Governance*, and ITAF to provide guidance and assistance to enterprises. COBIT 5 integrates all of this knowledge.

COBIT 5 Framework Integrator

Figure 10 provides a graphical description of how COBIT 5 achieves its role of an aligned and integrated framework.

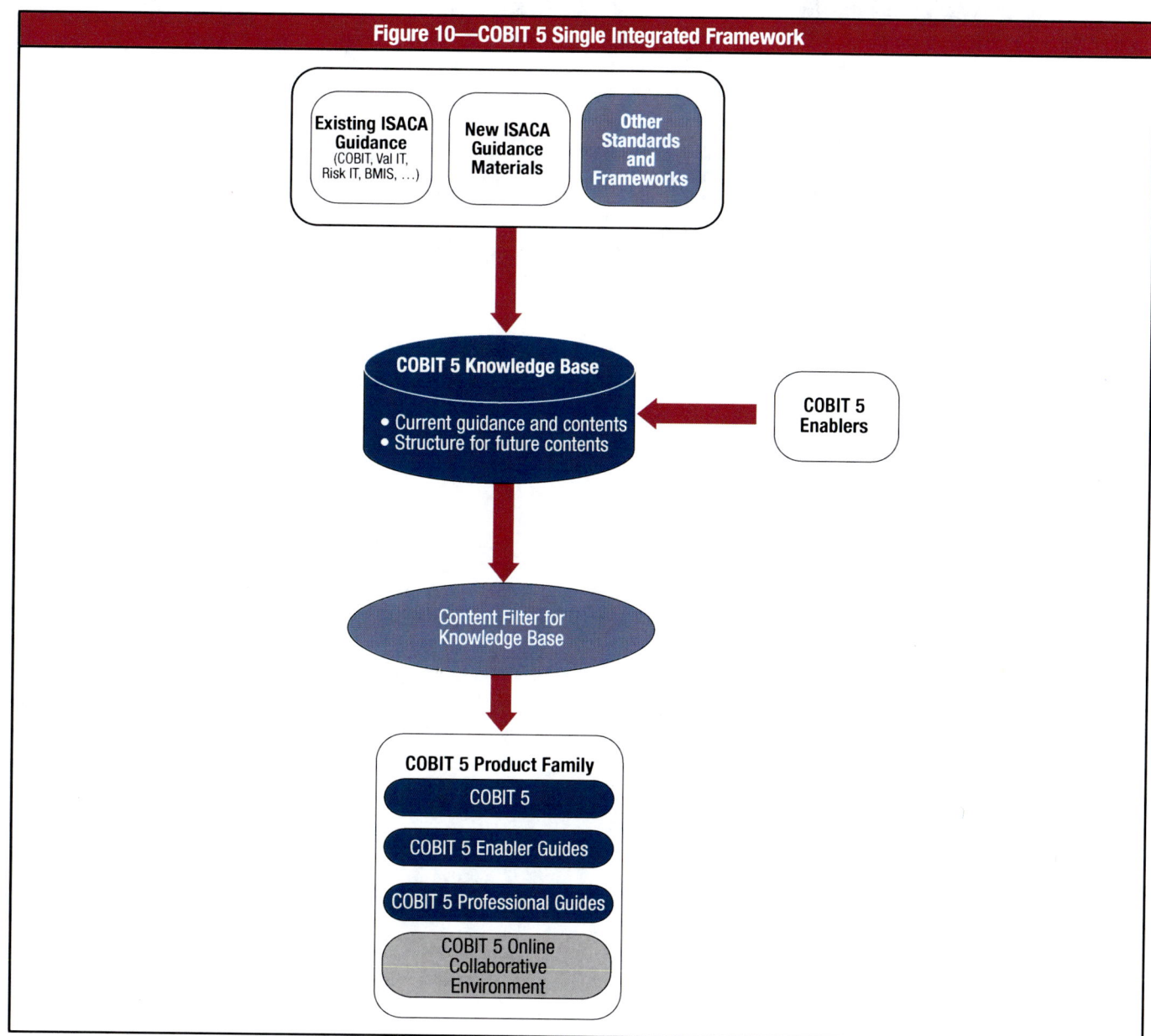

The COBIT 5 framework delivers to its stakeholders the most complete and up-to-date guidance (see **figure 11**) on governance and management of enterprise IT by:

• Researching and using a set of sources that have driven the new content development, including:
  – Bringing together the existing ISACA guidance (COBIT 4.1, Val IT 2.0, Risk IT, BMIS) into this single framework
  – Complementing this content with areas needing further elaboration and updates
  – Aligning to other relevant standards and frameworks, such as ITIL, TOGAF and ISO standards. A full list of references can be found in appendix A.
• Defining a set of governance and management enablers, which provide a structure for all guidance materials
• Populating a COBIT 5 knowledge base that contains all guidance and content produced now and will provide a structure for additional future content
• Providing a sound and comprehensive reference base of good practices

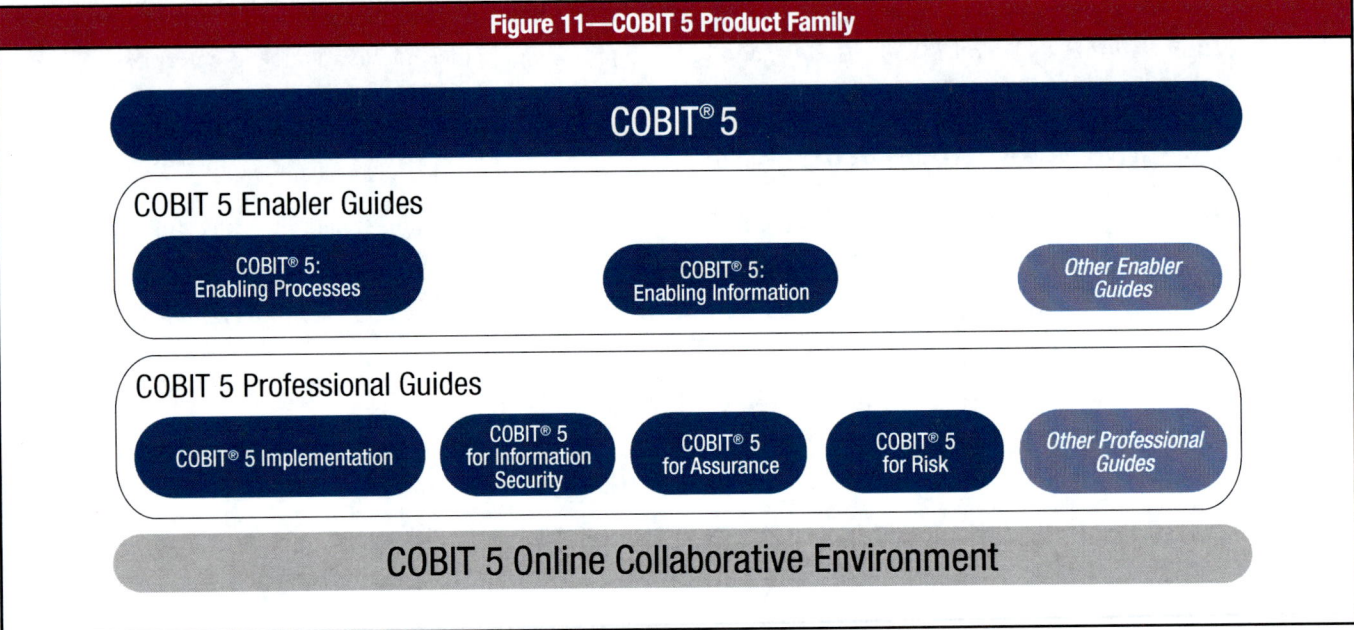

Figure 11—COBIT 5 Product Family

CHAPTER 5
PRINCIPLE 4: ENABLING A HOLISTIC APPROACH

COBIT 5 Enablers

Enablers are factors that, individually and collectively, influence whether something will work—in this case, governance and management over enterprise IT. Enablers are driven by the goals cascade, i.e., higher-level IT-related goals define what the different enablers should achieve.

The COBIT 5 framework describes seven categories of enablers (**figure 12**):
- **Principles, policies and frameworks** are the vehicle to translate the desired behaviour into practical guidance for day-to-day management.
- **Processes** describe an organised set of practices and activities to achieve certain objectives and produce a set of outputs in support of achieving overall IT-related goals.
- **Organisational structures** are the key decision-making entities in an enterprise.
- **Culture, ethics and behaviour** of individuals and of the enterprise are very often underestimated as a success factor in governance and management activities.
- **Information** is pervasive throughout any organisation and includes all information produced and used by the enterprise. Information is required for keeping the organisation running and well governed, but at the operational level, information is very often the key product of the enterprise itself.
- **Services, infrastructure and applications** include the infrastructure, technology and applications that provide the enterprise with information technology processing and services.
- **People, skills and competencies** are linked to people and are required for successful completion of all activities and for making correct decisions and taking corrective actions.

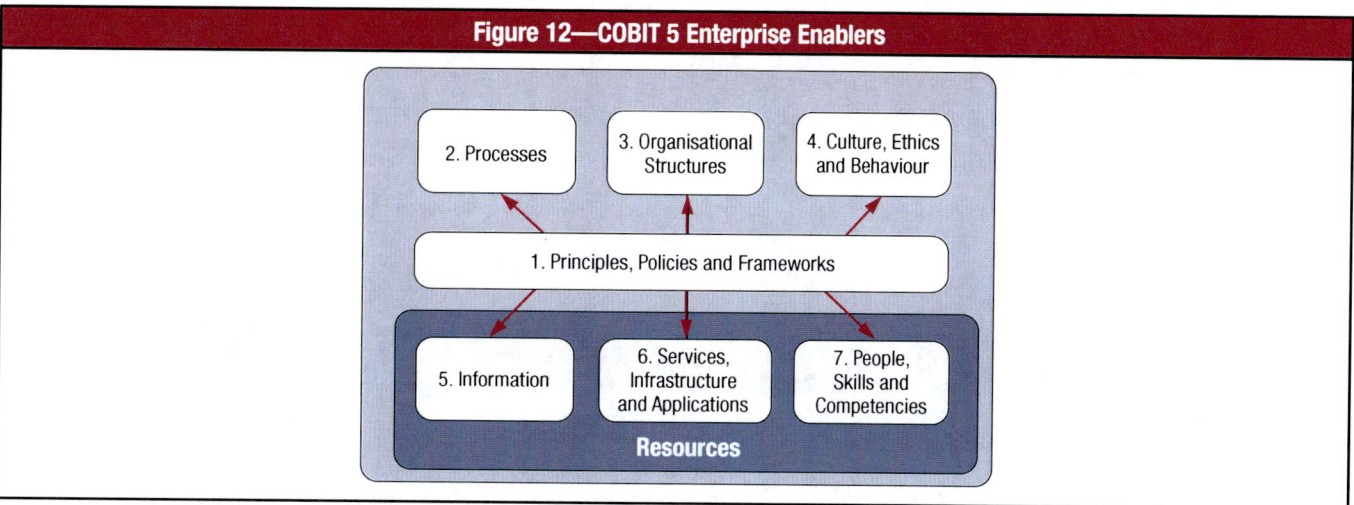

Figure 12—COBIT 5 Enterprise Enablers

Some of the enablers defined previously are also enterprise resources that need to be managed and governed as well. This applies to:
- Information, which needs to be managed as a resource. Some information, such as management reports and business intelligence information, are important enablers for the governance and management of the enterprise.
- Service, infrastructure and applications
- People, skills and competencies

Systemic Governance and Management Through Interconnected Enablers

Figure 12 also conveys the mindset that should be adopted for enterprise governance, including governance of IT, which is to achieve the main objectives of the enterprise. Any enterprise must always consider an interconnected set of enablers. That is, each enabler:
- Needs the input of other enablers to be fully effective, e.g., processes need information, organisational structures need skills and behaviour.
- Delivers output to the benefit of other enablers, e.g., processes deliver information, skills and behaviour make processes efficient.

So when dealing with governance and management of enterprise IT, good decisions can be taken only when this systemic nature of governance and management arrangements is taken into account. This means that to deal with any stakeholder need, all interrelated enablers have to be analysed for relevance and addressed if required. This mindset has to be driven by the top of the enterprise, as illustrated by the following examples.

EXAMPLE 3—GOVERNANCE AND MANAGEMENT OF ENTERPRISE IT
Providing operational IT services to all users requires service capabilities (infrastructure, application), for which people with the appropriate skill set and behaviour are required. A number of service delivery processes need to be implemented as well, supported by the appropriate organisational structures, showing how all enablers are required for successful service delivery.

EXAMPLE 4—GOVERNANCE AND MANAGEMENT OF ENTERPRISE IT
The need for information security requires a number of policies and procedures to be created and put in place. These policies, in turn, require a number of security-related practices to be implemented. However, if the enterprise's and personnel's culture and ethics are not appropriate, information security processes and procedures will not be effective.

COBIT 5 Enabler Dimensions

All enablers have a set of common dimensions. This set of common dimensions (**figure 13**):
• Provides a common, simple and structured way to deal with enablers
• Allows an entity to manage its complex interactions
• Facilitates successful outcomes of the enablers

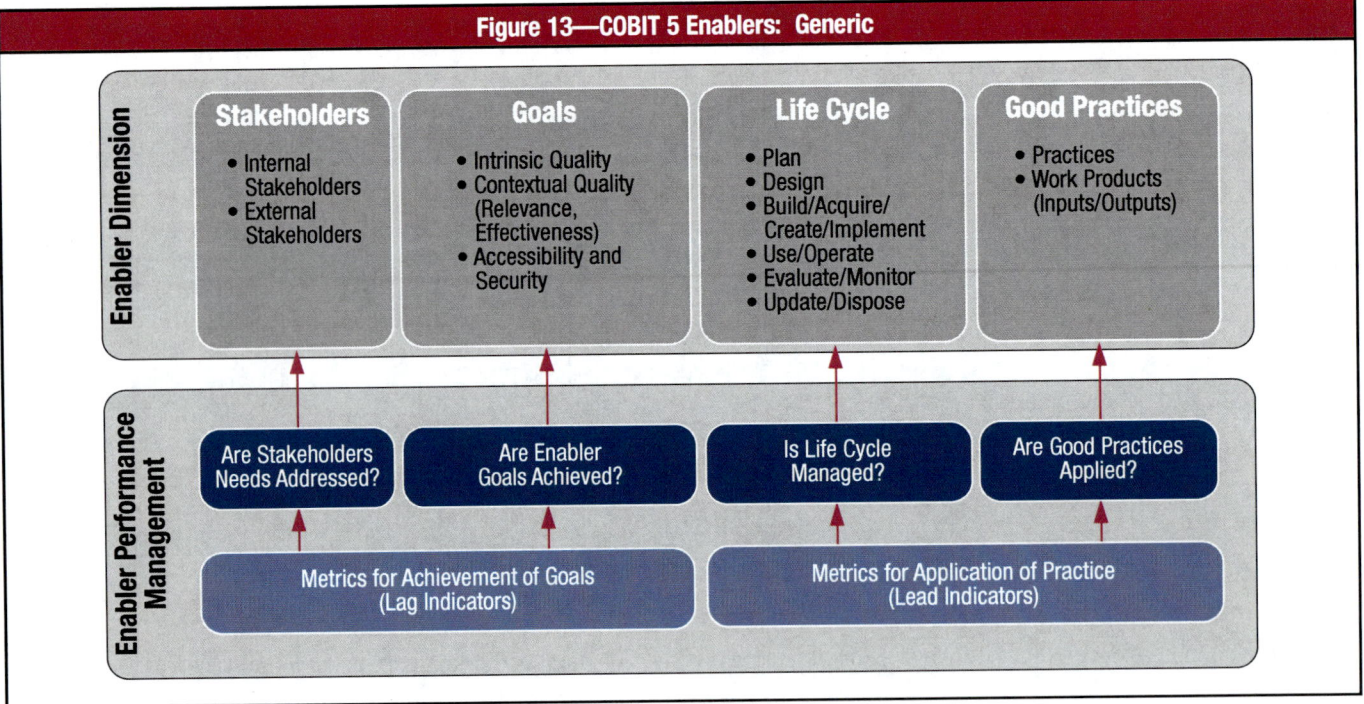

Figure 13—COBIT 5 Enablers: Generic

Enabler Dimensions

The four common dimensions for enablers are:
• **Stakeholders**—Each enabler has stakeholders (parties who play an active role and/or have an interest in the enabler). For example, processes have different parties who execute process activities and/or who have an interest in the process outcomes; organisational structures have stakeholders, each with his/her own roles and interests, that are part of the structures. Stakeholders can be internal or external to the enterprise, all having their own, sometimes conflicting, interests and needs. Stakeholders' needs translate to enterprise goals, which in turn translate to IT-related goals for the enterprise. A list of stakeholders is shown in **figure 7**.
• **Goals**—Each enabler has a number of goals, and enablers provide value by the achievement of these goals. Goals can be defined in terms of:
– Expected outcomes of the enabler
– Application or operation of the enabler itself

The enabler goals are the final step in the COBIT 5 goals cascade. Goals can be further split up in different categories:

– **Intrinsic quality**—The extent to which enablers work accurately, objectively and provide accurate, objective and reputable results

– **Contextual quality**—The extent to which enablers and their outcomes are fit for purpose given the context in which they operate. For example, outcomes should be relevant, complete, current, appropriate, consistent, understandable and easy to use.

– **Access and security**—The extent to which enablers and their outcomes are accessible and secured, such as:
  - Enablers are available when, and if, needed.
  - Outcomes are secured, i.e., access is restricted to those entitled and needing it.

• **Life cycle**—Each enabler has a life cycle, from inception through an operational/useful life until disposal. This applies to information, structures, processes, policies, etc. The phases of the life cycle consist of:
  – Plan (includes concepts development and concepts selection)
  – Design
  – Build/acquire/create/implement
  – Use/operate
  – Evaluate/monitor
  – Update/dispose

• **Good practices**—For each of the enablers, good practices can be defined. Good practices support the achievement of the enabler goals. Good practices provide examples or suggestions on how best to implement the enabler, and what work products or inputs and outputs are required. COBIT 5 provides examples of good practices for some enablers provided by COBIT 5 (e.g., processes). For other enablers, guidance from other standards, frameworks, etc., can be used.

Enabler Performance Management

Enterprises expect positive outcomes from the application and use of enablers. To manage performance of the enablers, the following questions will have to be monitored and thereby subsequently answered—based on metrics—on a regular basis:
• Are stakeholder needs addressed?
• Are enabler goals achieved?
• Is the enabler life cycle managed?
• Are good practices applied?

The first two bullets deal with the actual outcome of the enabler. The metrics used to measure to what extent the goals are achieved can be called 'lag indicators'.

The last two bullets deal with the actual functioning of the enabler itself, and metrics for this can be called 'lead indicators'.

Example of Enablers in Practice

Example 5 illustrates the enablers, their interconnections and the enabler dimensions, and how to use them for practical benefit.

EXAMPLE 5—ENABLERS

An organisation has appointed 'process managers' for IT-related processes, charged with defining and operating effective and efficient IT-related processes, in the context of good governance and management of enterprise IT.

Initially, the process managers will focus on the process enabler, considering the enabler dimensions:

• **Stakeholders:** Process stakeholders include all process actors, i.e., all parties that are responsible, accountable, consulted or informed (RACI) for, or during, process activities. For this, a RACI chart as described in *COBIT 5: Enabling Processes* can be used.

• **Goals:** For each process, adequate goals and related metrics need to be defined. For example, for a process *Manage Relationships* (process APO08 in *COBIT 5: Enabling Processes*) one can find a set of process goals and metrics such as:
  – **Goal:** Business strategies, plans and requirements are well understood, documented and approved.
    • **Metric:** Percent of programmes aligned with enterprise business requirements/priorities
  – **Goal:** Good relationships exist between the enterprise and the IT department.
    • **Metric:** Ratings of user and IT personnel satisfaction surveys

• **Life cycle:** Each process has a life cycle, i.e., it has to be created, executed and monitored, and adjusted when required. Eventually, processes cease to exist. In this case, the process managers would need to design and define the process first. They can use several elements from *COBIT 5: Enabling Processes* to design processes, i.e., to define responsibilities and to break the process down into practices and activities, and define process work products (inputs and outputs). In a later stage, the process needs to be made more robust and efficient, and for that purpose the process managers can raise the capability level of the process. The ISO/IEC 15504-inspired COBIT 5 Process Capability Model and the process capability attributes can be used for that purpose.

EXAMPLE 5—ENABLERS *(cont.)*

• **Good practice:** COBIT 5 describes in ample detail good practices for processes in *COBIT 5: Enabling Processes*, as mentioned under the previous point. Inspiration and example processes can be found there, covering the full spectrum of required activities for good governance and management of enterprise IT.
 	
 In addition to the guidance on the process enabler, the process managers can decide to look at a number of other enablers such as:
 • The RACI charts, which describe roles and responsibilities. Other enablers allow one to drill down on this dimension such as:
   – In the skills and competencies enabler, the required skills and competencies for each role can be defined, and appropriate goals (e.g., technical and behavioural skill levels) and associated metrics can be defined.
   – The RACI chart also contains a number of organisational structures. These structures can be further elaborated in the organisational structures enabler, where a more detailed description of the structure can be provided, expected outcomes and related metrics can be defined (e.g., decisions) and good practices can be defined (e.g., span of control, operating principles of the structure, level of authority).
 • Principles and policies will formalise the processes and prescribe why the process exists, to whom it is applicable and how the process is to be used. This is the focus area of the principles and policies enabler.

In appendix G, the seven categories of enablers are discussed in more detail. Reading this appendix is recommended for better understanding the enablers and how powerful they can be in organising governance and management of enterprise IT.

CHAPTER 6
PRINCIPLE 5: SEPARATING GOVERNANCE FROM MANAGEMENT

Governance and Management

The COBIT 5 framework makes a clear distinction between governance and management. These two disciplines encompass different types of activities, require different organisational structures and serve different purposes. The COBIT 5 view on this key distinction between governance and management is:
• **Governance**

> **Governance ensures that stakeholder needs, conditions and options are evaluated to determine balanced, agreed-on enterprise objectives to be achieved; setting direction through prioritisation and decision making; and monitoring performance and compliance against agreed-on direction and objectives.**

In most enterprises, governance is the responsibility of the board of directors under the leadership of the chairperson.

• **Management**

> **Management plans, builds, runs and monitors activities in alignment with the direction set by the governance body to achieve the enterprise objectives.**

In most enterprises, management is the responsibility of the executive management under the leadership of the CEO.

Interactions Between Governance and Management

From the definitions of governance and management, it is clear that they comprise different types of activities, with different responsibilities; however, given the role of governance—to evaluate, direct and monitor—a set of interactions is required between governance and management to result in an efficient and effective governance system. These interactions, using the enabler structure, are shown at a high level in **figure 14**.

Figure 14—COBIT 5 Governance and Management Interactions	
Enabler	**Governance-Management Interaction**
Processes	In the illustrative COBIT 5 process model (*COBIT 5: Enabling Processes*), a distinction is made between governance and management processes, including specific sets of practices and activities for each. The process model also includes RACI charts, describing the responsibilities of different organisational structures and roles within the enterprise.
Information	The process model describes inputs to and outputs from the different process practices to other processes, including information exchanged between governance and management processes. Information used for evaluating, directing and monitoring enterprise IT is exchanged between governance and management as described in the process model inputs and outputs.
Organisational structures	A number of organisational structures are defined in each enterprise; structures can sit in the governance space or the management space, depending on their composition and scope of decisions. Because governance is about setting the direction, interaction takes place between the decisions taken by the governance structures—e.g., deciding about the investment portfolio and setting risk appetite—and the decisions and operations implementing the former.
Principles, policies and frameworks	Principles, policies and frameworks are the vehicle by which governance decisions are institutionalised within the enterprise, and for that reason are an interaction between governance decisions (direction setting) and management (execution of decisions).
Culture, ethics and behaviour	Behaviour is also a key enabler of good governance and management of the enterprise. It is set at the top—leading by example—and is therefore an important interaction between governance and management.
People, skills and competencies	Governance and management activities require different skill sets, but an essential skill for both governance body members and management is to understand both tasks and how they are different.
Services, infrastructure and applications	Services are required, supported by applications and infrastructure to provide the governance body with adequate information and to support the governance activities of evaluating, setting direction and monitoring.

COBIT 5 Process Reference Model

COBIT 5 is not prescriptive, but it advocates that enterprises implement governance and management processes such that the key areas are covered, as shown in **figure 15**.

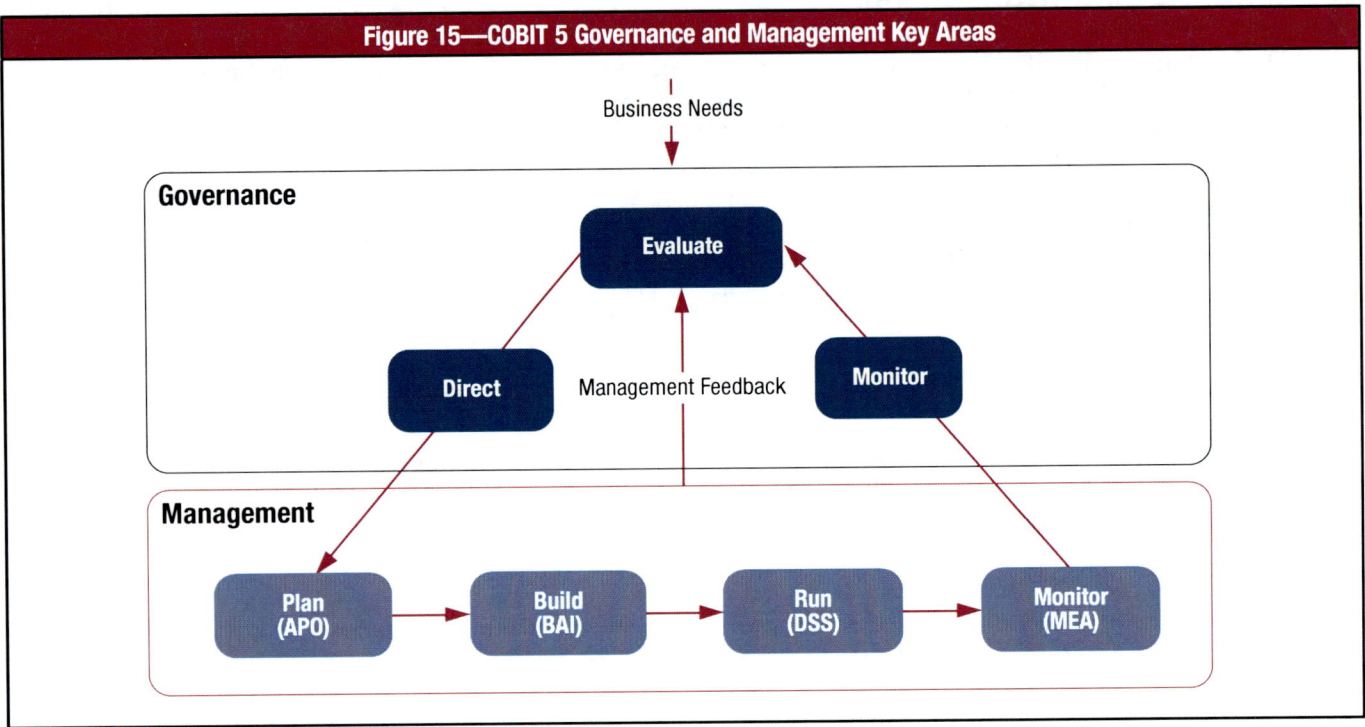

An enterprise can organise its processes as it sees fit, as long as all necessary governance and management objectives are covered. Smaller enterprises may have fewer processes; larger and more complex enterprises may have many processes, all to cover the same objectives.

COBIT 5 includes a process reference model, which defines and describes in detail a number of governance and management processes. It represents all of the processes normally found in an enterprise relating to IT activities, providing a common reference model understandable to operational IT and business managers. The proposed process model is a complete, comprehensive model, but it is not the only possible process model. Each enterprise must define its own process set, taking into account its specific situation.

Incorporating an operational model and a common language for all parts of the enterprise involved in IT activities is one of the most important and critical steps towards good governance. It also provides a framework for measuring and monitoring IT performance, providing IT assurance, communicating with service providers, and integrating best management practices.

The COBIT 5 process reference model divides the governance and management processes of enterprise IT into two main process domains:
• **Governance**—Contains five governance processes; within each process, evaluate, direct and monitor (EDM)[5] practices are defined.
• **Management**—Contains four domains, in line with the responsibility areas of plan, build, run and monitor (PBRM), and provides end-to-end coverage of IT. These domains are an evolution of the COBIT 4.1 domain and process structure. The names of the domains are chosen in line with these main area designations, but contain more verbs to describe them:
  – Align, Plan and Organise (APO)
  – Build, Acquire and Implement (BAI)
  – Deliver, Service and Support (DSS)
  – Monitor, Evaluate and Assess (MEA)

---

[5] In the context of the governance domain, 'monitoring' means those activities where the governance body checks to what extent the direction that has been set for management is actually applied.

Each domain contains a number of processes. Although, as described previously, most of the processes require 'planning', 'implementation', 'execution' and 'monitoring' activities within the process or within the specific issue being addressed (e.g., quality, security), they are placed in domains in line with what is generally the most relevant area of activity when looking at IT at the enterprise level.

The COBIT 5 process reference model is the successor of the COBIT 4.1 process model, with the Risk IT and Val IT process models integrated as well.

Figure 16 shows the complete set of 37 governance and management processes within COBIT 5. The details of all processes, according to the process model described previously, are included in *COBIT 5: Enabling Processes.*

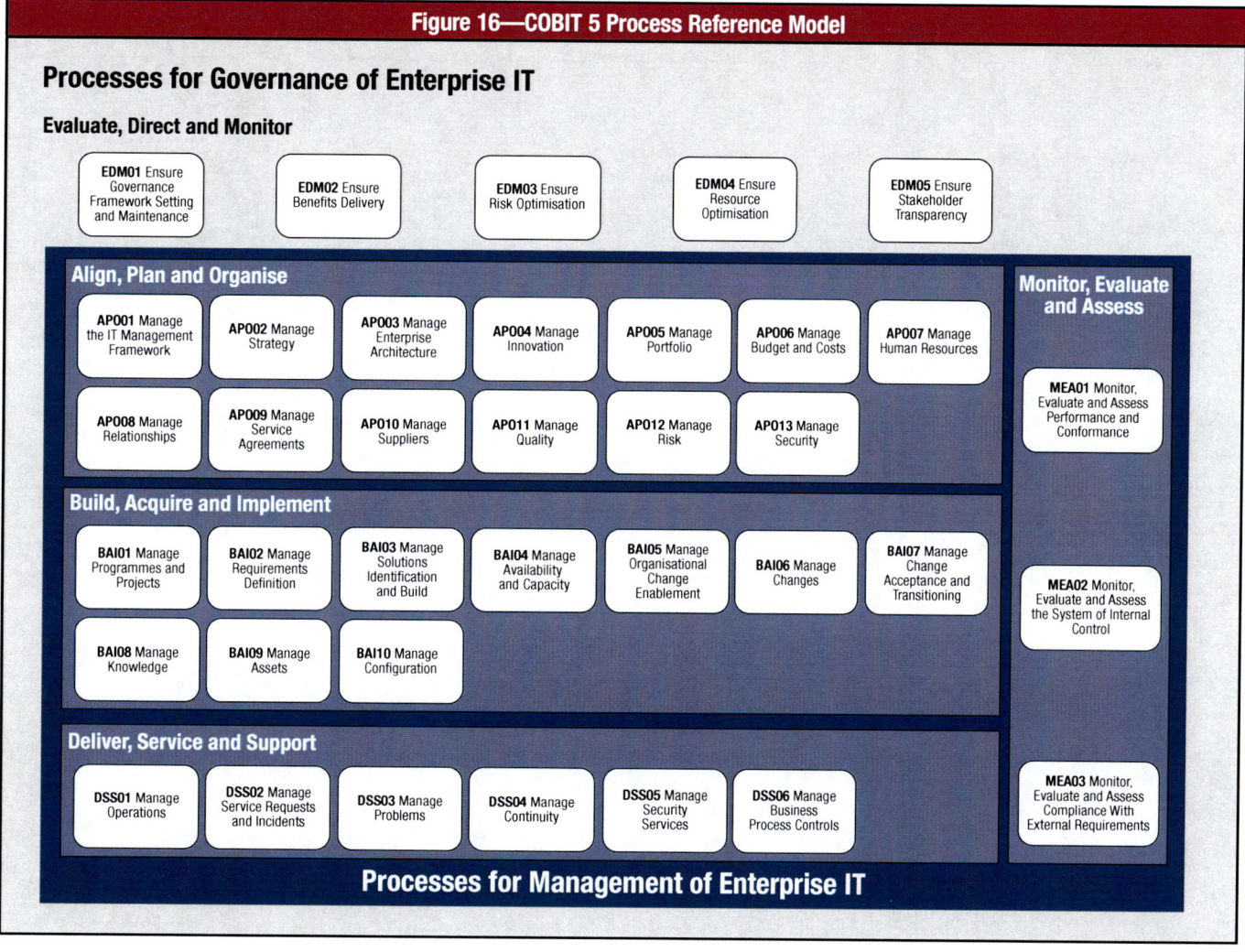

Figure 16—COBIT 5 Process Reference Model

Page intentionally left blank

CHAPTER 7
IMPLEMENTATION GUIDANCE

Introduction

Optimal value can be realised from leveraging COBIT only if it is effectively adopted and adapted to suit each enterprise's unique environment. Each implementation approach will also need to address specific challenges, including managing changes to culture and behaviour.

ISACA provides practical and extensive implementation guidance in its publication *COBIT 5 Implementation*,[6] which is based on a continual improvement life cycle. It is not intended to be a prescriptive approach nor a complete solution, but rather a guide to avoid commonly encountered pitfalls, leverage good practices and assist in the creation of successful outcomes. The guide is also supported by an implementation tool kit containing a variety of resources that will be continually enhanced. Its content includes:
• Self-assessment, measurement and diagnostic tools
• Presentations aimed at various audiences
• Related articles and further explanations

The purpose of this chapter is to introduce the implementation and continual improvement life cycle at a high level and to highlight a number of important topics from *COBIT 5 Implementation* such as:
• Making a business case for the implementation and improvement of the governance and management of IT
• Recognising typical pain points and trigger events
• Creating the appropriate environment for implementation
• Leveraging COBIT to identify gaps and guide the development of enablers such as policies, processes, principles, organisational structures, and roles and responsibilities

Considering the Enterprise Context

The governance and management of enterprise IT do not occur in a vacuum. Every enterprise needs to design its own implementation plan or road map, depending on factors in the enterprise's specific internal and external environment such as the enterprise's:
• Ethics and culture
• Applicable laws, regulations and policies
• Mission, vision and values
• Governance policies and practices
• Business plan and strategic intentions
• Operating model and level of maturity
• Management style
• Risk appetite
• Capabilities and available resources
• Industry practices

It is equally important to leverage and build on existing enterprise governance enablers.

The optimal approach for the governance and management of enterprise IT will be different for every enterprise, and the context needs to be understood and considered to adopt and adapt COBIT effectively in the implementation of governance and management of enterprise IT enablers. COBIT is often underpinned with other frameworks, good practices and standards, and these, too, need to be adapted to suit specific requirements.

Key success factors for successful implementation include:
• Top management providing the direction and mandate for the initiative, as well as visible ongoing commitment and support
• All parties supporting the governance and management processes to understand the business and IT objectives
• Ensuring effective communication and enablement of the necessary changes
• Tailoring COBIT and other supporting good practices and standards to fit the unique context of the enterprise
• Focussing on quick wins and prioritising the most beneficial improvements that are easiest to implement

---

[6] *www.isaca.org/cobit*

Creating the Appropriate Environment

It is important for implementation initiatives leveraging COBIT to be properly governed and adequately managed. Major IT-related initiatives often fail due to inadequate direction, support and oversight by the various required stakeholders, and the implementation of governance or management of IT enablers leveraging COBIT is no different. Support and direction from key stakeholders are critical so that improvements are adopted and sustained. In a weak enterprise environment (such as an unclear overall business operating model or lack of enterprise-level governance enablers), this support and participation are even more important.

Enablers leveraging COBIT should provide a solution addressing real business needs and issues rather than serving as ends in themselves. Requirements based on current pain points and drivers should be identified and accepted by management as areas that need to be addressed. High-level health checks, diagnostics or capability assessments based on COBIT are excellent tools to raise awareness, create consensus and generate a commitment to act. The commitment and buy-in of the relevant stakeholders need to be solicited from the beginning. To achieve this, implementation objectives and benefits need to be clearly expressed in business terms and summarised in a business case outline.

Once commitment has been obtained, adequate resources need to be provided to support the programme. Key programme roles and responsibilities should be defined and assigned. Care should be taken on an ongoing basis to maintain commitment from all affected stakeholders.

Appropriate structures and processes for oversight and direction should be established and maintained. These structures and processes should also ensure ongoing alignment with enterprisewide governance and risk management approaches.

Visible support and commitment should be provided by key stakeholders such as the board and executives to set the 'tone at the top' and ensure commitment for the programme at all levels.

Recognising Pain Points and Trigger Events

There are a number of factors that may indicate a need for improved governance and management of enterprise IT.

By using pain points or trigger events as the launching point for implementation initiatives, the business case for governance or management of enterprise IT improvement can be related to practical, everyday issues being experienced. This will improve buy-in and create the sense of urgency within the enterprise that is necessary to kick off the implementation. In addition, quick wins can be identified and value-add can be demonstrated in those areas that are the most visible or recognisable in the enterprise. This provides a platform for introducing further changes and can assist in gaining widespread senior management commitment and support for more pervasive changes.

Examples of some of the typical pain points for which new or revised governance or management of IT enablers can be a solution (or part of a solution), as identified in *COBIT 5 Implementation*, are:
• Business frustration with failed initiatives, rising IT costs and a perception of low business value
• Significant incidents related to IT risk, such as data loss or project failure
• Outsourcing service delivery problems, such as consistent failure to meet agreed-on service levels
• Failure to meet regulatory or contractual requirements
• IT limiting the enterprise's innovation capabilities and business agility
• Regular audit findings about poor IT performance or reported IT quality of service problems
• Hidden and rogue IT spending
• Duplication or overlap between initiatives or wasting resources, such as premature project termination
• Insufficient IT resources, staff with inadequate skills or staff burnout/dissatisfaction
• IT-enabled changes failing to meet business needs and delivered late or over budget
• Board members, executives or senior managers who are reluctant to engage with IT, or a lack of committed and satisfied business sponsors for IT
• Complex IT operating models

In addition to these pain points, other events in the enterprise's internal and external environment can signal or trigger a focus on the governance and management of IT. Examples from chapter 3 in the *COBIT 5 Implementation* publication are:
• Merger, acquisition or divestiture
• A shift in the market, economy or competitive position
• A change in the business operating model or sourcing arrangements
• New regulatory or compliance requirements

- A significant technology change or paradigm shift
- An enterprisewide governance focus or project
- A new CEO, CFO, CIO, etc.
- External audit or consultant assessments
- A new business strategy or priority

Enabling Change

Successful implementation depends on implementing the appropriate change (the appropriate governance or management enablers) in the appropriate way. In many enterprises, there is a significant focus on the first aspect—core governance or management of IT—but not enough emphasis on managing the human, behavioural and cultural aspects of the change and motivating stakeholders to buy into the change.

It should not be assumed that the various stakeholders involved in, or impacted by, new or revised enablers will readily accept and adopt the change. The possibility of ignorance and/or resistance to change needs to be addressed through a structured and proactive approach. Also, optimal awareness of the implementation programme should be achieved through a communication plan that defines what will be communicated, in what way and by whom, throughout the various phases of the programme.

Sustainable improvement can be achieved either by gaining the commitment of the stakeholders (investment in winning hearts and minds, the leaders' time, and in communicating and responding to the workforce) or, where still required, by enforcing compliance (investment in processes to administer, monitor and enforce). In other words, human, behavioural and cultural barriers need to be overcome so that there is a common interest to properly adopt change, instil a will to adopt change, and to ensure the ability to adopt change.

A Life Cycle Approach

The implementation life cycle provides a way for enterprises to use COBIT to address the complexity and challenges typically encountered during implementations. The three interrelated components of the life cycle are the:
1. Core continual improvement life cycle—This is not a one-off project.
2. Enablement of change—Addressing the behavioural and cultural aspects
3. Management of the programme

As discussed previously, the appropriate environment needs to be created to ensure the success of the implementation or improvement initiative. The life cycle and its seven phases are illustrated in **figure 17**.

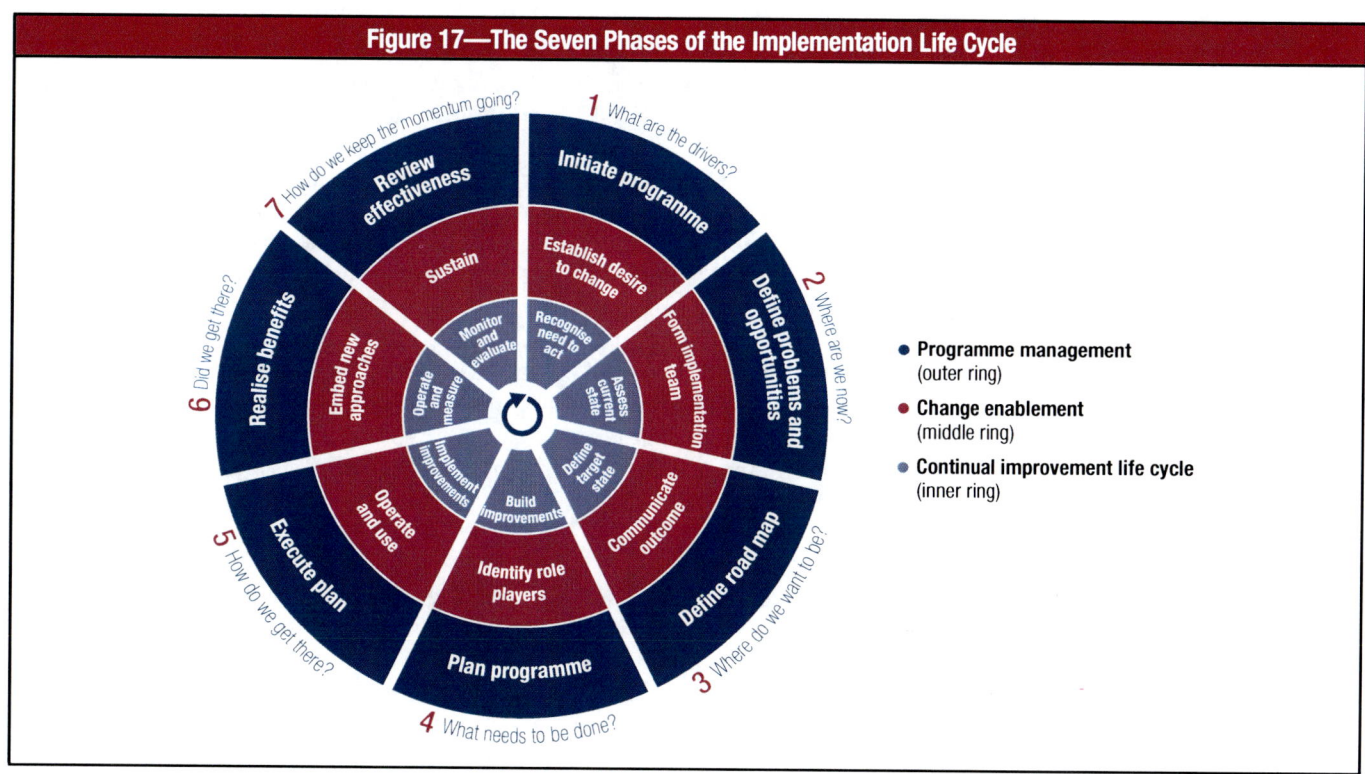

Figure 17—The Seven Phases of the Implementation Life Cycle

- Programme management (outer ring)
- Change enablement (middle ring)
- Continual improvement life cycle (inner ring)

Phase 1 starts with recognising and agreeing to the need for an implementation or improvement initiative. It identifies the current pain points and triggers and creates a desire to change at executive management levels.

Phase 2 is focused on defining the scope of the implementation or improvement initiative using COBIT's mapping of enterprise goals to IT-related goals to the associated IT processes, and considering how risk scenarios could also highlight key processes on which to focus. High-level diagnostics can also be useful for scoping and understanding high-priority areas on which to focus. An assessment of the current state is then performed, and issues or deficiencies are identified by carrying out a process capability assessment. Large-scale initiatives should be structured as multiple iterations of the life cycle—for any implementation initiative exceeding six months there is a risk of losing momentum, focus and buy-in from stakeholders.

During **phase 3**, an improvement target is set, followed by a more detailed analysis leveraging COBIT's guidance to identify gaps and potential solutions. Some solutions may be quick wins and others more challenging and longer-term activities. Priority should be given to initiatives that are easier to achieve and those likely to yield the greatest benefits.

Phase 4 plans practical solutions by defining projects supported by justifiable business cases. A change plan for implementation is also developed. A well-developed business case helps to ensure that the project's benefits are identified and monitored.

The proposed solutions are implemented into day-to-day practices in **phase 5**. Measures can be defined and monitoring established, using COBIT's goals and metrics to ensure that business alignment is achieved and maintained and performance can be measured. Success requires the engagement and demonstrated commitment of top management as well as ownership by the affected business and IT stakeholders.

Phase 6 focuses on the sustainable operation of the new or improved enablers and the monitoring of the achievement of expected benefits.

During **phase 7**, the overall success of the initiative is reviewed, further requirements for the governance or management of enterprise IT are identified, and the need for continual improvement is reinforced.

Over time, the life cycle should be followed iteratively while building a sustainable approach to the governance and management of enterprise IT.

Getting Started: Making the Business Case

To ensure the success of implementation initiatives leveraging COBIT, the need to act should be widely recognised and communicated within the enterprise. This can be in the form of a 'wake-up call' (where specific pain points are being experienced, as discussed previously) or an expression of the improvement opportunity to be pursued and, very important, the benefits that will be realised. An appropriate level of urgency needs to be instilled and the key stakeholders should be aware of the risk of not taking action as well as the benefits of undertaking the programme.

The initiative should be owned by a sponsor, involve all key stakeholders and be based on a business case. Initially, this can be at a high level from a strategic perspective—from the top down—starting with a clear understanding of the desired business outcomes and progressing to a detailed description of critical tasks and milestones as well as key roles and responsibilities. The business case is a valuable tool available to management in guiding the creation of business value. At a minimum, the business case should include the following:
- The business benefits targeted, their alignment with business strategy and the associated benefit owners (who in the business will be responsible for securing them). This could be based on pain points and trigger events.
- The business changes needed to create the envisioned value. This could be based on health checks and capability gap analyses and should clearly state both what is in scope and what is out of scope.
- The investments needed to make the governance and management of enterprise IT changes (based on estimates of projects required)
- The ongoing IT and business costs
- The expected benefits of operating in the changed way
- The risk inherent in the previous bullets, including any constraints or dependencies (based on challenges and success factors)
- Roles, responsibilities and accountabilities related to the initiative
- How the investment and value creation will be monitored throughout the economic life cycle, and the metrics to be used (based on goals and metrics)

The business case is not a one-time static document, but a dynamic operational tool that must be continually updated to reflect the current view of the future so that a view of the viability of the programme can be maintained.

It can be difficult to quantify the benefits of implementation or improvement initiatives, and care should be taken to commit only to benefits that are realistic and achievable. Studies conducted across a number of enterprises could provide useful information on benefits that have been achieved.

EXAMPLE 6—GOVERNANCE OF IT STATISTICS
ITGI commissioned a market research project on the governance of IT[7] by PwC, with more than 800 IT and business respondents in 21 countries. Thirty-eight percent of respondents cited lower IT costs as an outcome of governance of IT practices, 28.1 percent cited improved business competitiveness and 27.1 percent indicated an improved return on IT investments. In addition, a number of less tangible benefits were reported such as improved management of IT-related risk (42.2 percent of respondents), improved communication and relationships between business and IT (39.6 percent of respondents) and improved IT delivery of business objectives (37.3 percent of respondents).
ISACA has also undertaken research that[8] explores and demonstrates the business value of COBIT. The data set resulting from the research offers many analysis opportunities and clarifies the relationship between the enterprise governance of IT and business performance.
Another study conducted across 250 enterprises worldwide found that those enterprises with superior IT governance had at least a 20 percent higher profitability than firms with poor governance, given the same objectives.[9] It argues that IT business value results directly from effective IT governance.
Finally, another research case in the airline industry concluded that the implementation and ongoing assurance of enterprise governance of IT restored trust between business and IT, and resulted in an increased alignment of investments to strategic goals. Also, more tangible benefits were reported in this case, including lowered IT continuity cost per business production unit, and the freeing up of funds for innovation. Other cross-case research in the financial sector demonstrated that organisations with better governance of IT approaches clearly obtained higher business/IT alignment maturity scores.[10]

[7] ITGI, *Global Status Report on the Governance of Enterprise IT (GEIT)—2011*, USA, 2011, *www.isaca.org/Knowledge-Center/Research/ResearchDeliverables/Pages/Global-Status-Report-on-the-Governance-of-Enterprise-IT-GEIT-2011.aspx*

[8] ISACA, *Building the Business Case for COBIT® and Val IT™ Executive Briefing*, USA, 2009, *www.isaca.org/Knowledge-Center/Research/ResearchDeliverables/Pages/Building-the-Business-Case-for-COBIT-and-Val-IT-Executive-Briefing.aspx*

[9] Weill, Peter; Jeanne W. Ross; *IT Governance: How Top Performers Manage IT Decision Rights for Superior Results*, Harvard Business School Press, USA, 2004

[10] De Haes, Steven; Dirk Gemke; John Thorp; Wim Van Grembergen; 'Analyzing IT Value Management @ KLM Through the Lens of Val IT', *ISACA Journal*, 2011, vol 4. Van Grembergen, Wim; Steven De Haes; *Enterprise Governance of IT: Achieving Alignment and Value*, Springer, USA, 2009

Page intentionally left blank

CHAPTER 8
THE COBIT 5 PROCESS CAPABILITY MODEL

Introduction

Users of COBIT 4.1, Risk IT and Val IT are familiar with the process maturity models included in those frameworks. These models are used to measure the current or 'as-is' maturity of an enterprise's IT-related processes, to define a required 'to-be' state of maturity, and to determine the gap between them and how to improve the process to achieve the desired maturity level.

The COBIT 5 product set includes a process capability model, based on the internationally recognised ISO/IEC 15504 Software Engineering—Process Assessment standard. This model will achieve the same overall objectives of process assessment and process improvement support, i.e., it will provide a means to measure the performance of any of the governance (EDM-based) processes or management (PBRM-based) processes, and will allow areas for improvement to be identified.

However, the new model is different from the COBIT 4.1 maturity model in its design and use, and for that reason, the following topics are discussed:
• Differences between the COBIT 5 and the COBIT 4.1 models
• Benefits of the COBIT 5 model
• Summary of the differences that COBIT 5 users will encounter in practice
• Performing a COBIT 5 capability assessment

Details of the COBIT 5 capability assessment approach are contained in the ISACA publication *COBIT® Process Assessment Model (PAM): Using COBIT® 4.1.*[11]

Although this approach will provide valuable information about the state of processes, processes are just one of the seven governance and management enablers. By consequence, process assessments will not provide the full picture on the state of governance of an enterprise. For that, the other enablers need to be assessed as well.

Differences Between the COBIT 4.1 Maturity Model and the COBIT 5 Process Capability Model

The elements of the COBIT 4.1 maturity model approach are shown in **figure 18**.

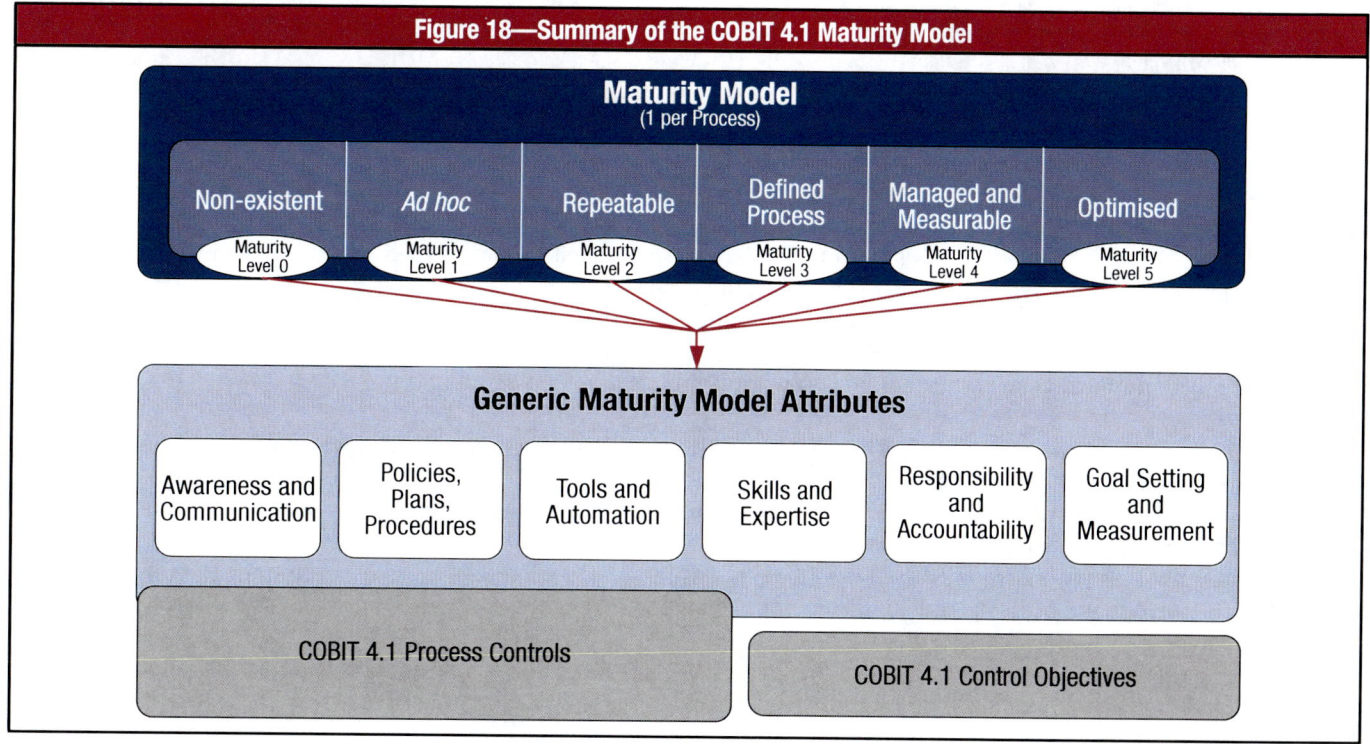

Figure 18—Summary of the COBIT 4.1 Maturity Model

---

[11] www.isaca.org/cobit-pam

Using the COBIT 4.1 maturity model for process improvement purposes—assessing a process maturity, defining a target maturity level and identifying the gaps—required using the following COBIT 4.1 components:
• First, an assessment needed to be made whether control objectives for the process were met.
• Next, the maturity model included in the management guideline for each process could be used to obtain a maturity profile of the process.
• In addition, the generic maturity model in COBIT 4.1 provided six distinct attributes that were applicable for each process and that assisted in obtaining a more detailed view on the processes' maturity level.
• Process controls are generic control objectives—they also needed to be reviewed when a process assessment was made. Process controls partially overlap with the generic maturity model attributes.

The COBIT 5 process capability approach can be summarised as shown in **figure 19**.

Figure 19—Summary of the COBIT 5 Process Capability Model

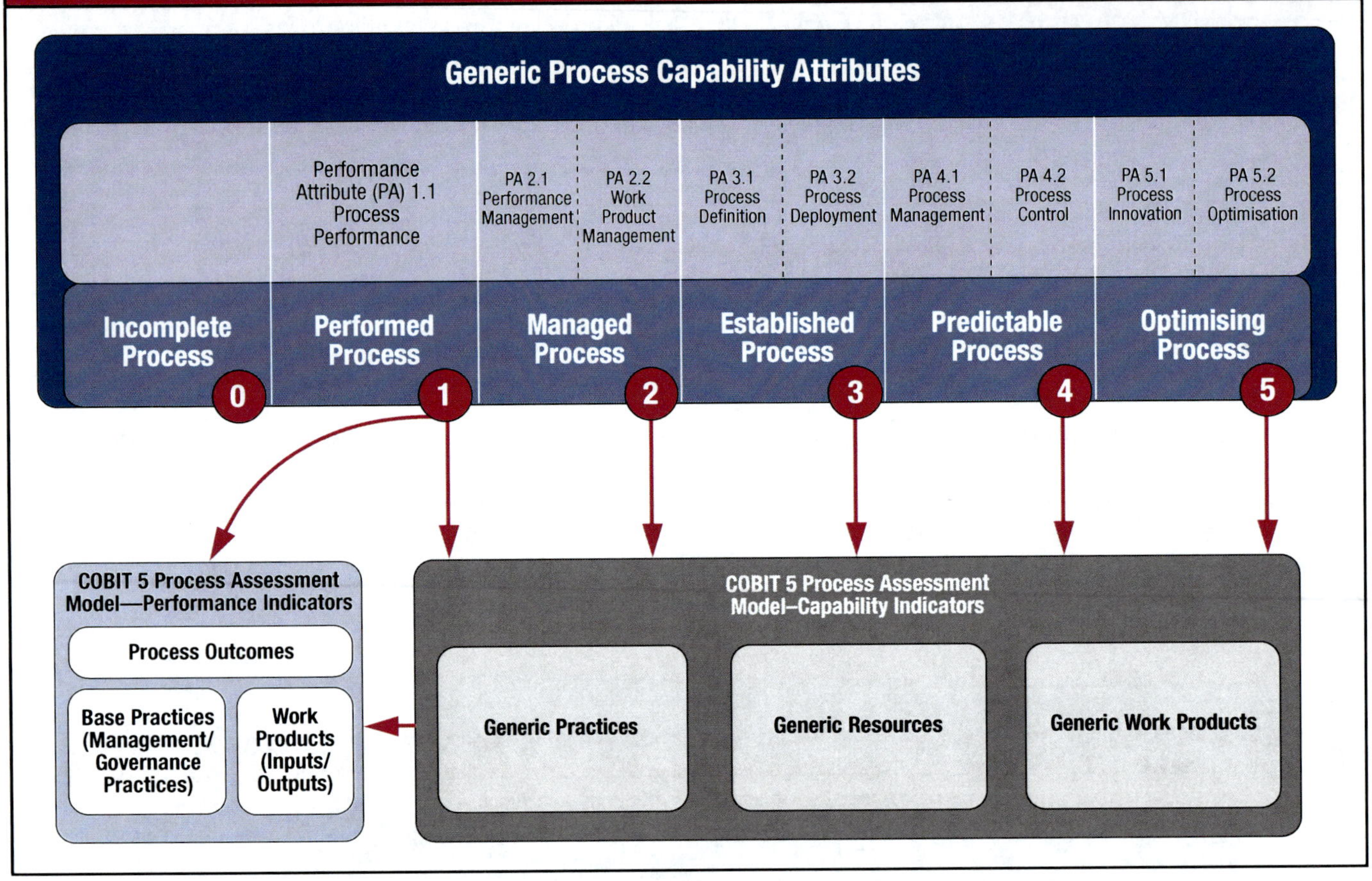

Generic Process Capability Attributes

| Performance Attribute (PA) 1.1 Process Performance | PA 2.1 Performance Management | PA 2.2 Work Product Management | PA 3.1 Process Definition | PA 3.2 Process Deployment | PA 4.1 Process Management | PA 4.2 Process Control | PA 5.1 Process Innovation | PA 5.2 Process Optimisation |

| Incomplete Process | Performed Process | Managed Process | Established Process | Predictable Process | Optimising Process |
| 0 | 1 | 2 | 3 | 4 | 5 |

COBIT 5 Process Assessment Model—Performance Indicators

Process Outcomes

Base Practices (Management/ Governance Practices) — Work Products (Inputs/ Outputs)

COBIT 5 Process Assessment Model–Capability Indicators

Generic Practices — Generic Resources — Generic Work Products

There are six levels of capability that a process can achieve, including an 'incomplete process' designation if the practices in it do not achieve the intended purpose of the process:
• **0 Incomplete process**—The process is not implemented or fails to achieve its process purpose. At this level, there is little or no evidence of any systematic achievement of the process purpose.
• **1 Performed process** (one attribute)—The implemented process achieves its process purpose.
• **2 Managed process** (two attributes)—The previously described performed process is now implemented in a managed fashion (planned, monitored and adjusted) and its work products are appropriately established, controlled and maintained.
• **3 Established process** (two attributes)—The previously described managed process is now implemented using a defined process that is capable of achieving its process outcomes.
• **4 Predictable process** (two attributes)—The previously described established process now operates within defined limits to achieve its process outcomes.
• **5 Optimising process** (two attributes)—The previously described predictable process is continuously improved to meet relevant current and projected business goals.

Each capability level can be achieved only when the level below has been fully achieved. For example, a process capability level 3 (established process) requires the process definition and process deployment attributes to be largely achieved, on top of full achievement of the attributes for a process capability level 2 (managed process).

There is a significant distinction between process capability level 1 and the higher capability levels. Process capability level 1 achievement requires the process performance attribute to be largely achieved, which actually means that the process is being successfully performed and the required outcomes obtained by the enterprise. The higher capability levels then add different attributes to it. In this assessment scheme, achieving a capability level 1, even on a scale to 5, is already an important achievement for an enterprise. Note that each individual enterprise shall choose (based on cost-benefit and feasibility reasons) its target or desired level, which very seldom will happen to be one of the highest.

The most important differences between an ISO/IEC 15504-based process capability assessment and the current COBIT 4.1 maturity model (and the similar Val IT and Risk IT domain-based maturity models) can be summarised as follows:
• The naming and meaning of the ISO/IEC 15504-defined capability levels are quite different from the current COBIT 4.1 maturity levels for processes.
• In ISO/IEC 15504, capability levels are defined by a set of nine process attributes. These attributes cover some ground covered by the current COBIT 4.1 maturity attributes and/or process controls, but only to a certain extent and in a different way.

Requirements for an ISO/IEC 15504:2-compliant process reference model prescribe that in the description of any process that will be assessed, i.e., any COBIT 5 governance and/or management process:
• The process is described in terms of its purpose and outcomes.
• The process description shall not contain any aspects of the measurement framework beyond level 1, which means that any characteristic of a process attribute beyond level 1 cannot appear inside a process description. Whether a process is measured and monitored, or whether it is formally described, etc., cannot be part of a process description or any of the management practices/activities underneath. This means that the process descriptions—as included in *COBIT 5: Enabling Processes*—contain only the necessary steps to achieve the actual process purpose and goals.
• Following from the previous bullets, the common attributes applicable to all enterprise processes, which produced duplicative control objectives in the *COBIT® 3rd Edition* publication and were grouped into the process control (PC) objectives in COBIT 4.1, are now defined in levels 2 to 5 of the assessment model.

Differences in Practice[12]

From the previous descriptions, it is clear that there are some practical differences associated with the change in process assessment models. Users need to be aware of these changes and be prepared to take them into account in their action plans.

The main changes to be considered include:
• Although it is tempting to compare assessment results between COBIT 4.1 and COBIT 5 because of apparent similarities to the number scales and words used to describe them, such a comparison is difficult because of the differences in scope, focus and intent, as illustrated in **figure 20**.
• In general, scores will be lower with the COBIT 5 process capability model, as shown in **figure 20**. In the COBIT 4.1 maturity model, a process could achieve a level 1 or 2 without fully achieving all the process's objectives; in the COBIT 5 process capability level, this will result in a lower score of 0 or 1.

The COBIT 4.1 and COBIT 5 capability scales can be considered to 'map' approximately as shown in **figure 20.**
• There is no longer a specific maturity model per process included with the detailed process contents in COBIT 5 because the ISO/IEC 15504 process capability assessment approach does not require this and even prohibits this approach. Instead, the approach defines the information required in the 'process reference model' (the process model to be used for the assessment):
  – Process description, with the purpose statements
  – Base practices, which are the equivalent of process governance or management practices in COBIT 5 terms
  – Work products, which are the equivalent of the inputs and outputs in COBIT 5 terms
• The COBIT 4.1 maturity model produced a maturity profile of an enterprise. The main purpose of this profile was to identify in which dimensions or for which attributes there were specific weaknesses that needed improvement. This approach was used by enterprises when there was an improvement focus rather than a need to obtain one maturity number for reporting purposes. In COBIT 5 the assessment model provides a measurement scale for each capability attribute and guidance on how to apply it, so for each process an assessment can be made for each of the nine capability attributes.
• The maturity attributes in COBIT 4.1 and the COBIT 5 process capability attributes are not identical. They overlap/map to a certain extent, as shown in **figure 21**. Enterprises having used the maturity model attributes approach in COBIT 4.1 can reuse their existing assessment data and reclassify them under the COBIT 5 attribute assessments based on **figure 21**.

---

[12] More information on the new ISO/IEC 15504-based COBIT Assessment Programme can be found at *www.isaca.org/cobit-assessment-programme.*

Figure 20—Comparison Table of Maturity Levels (COBIT 4.1) and Process Capability Levels (COBIT 5)		
COBIT 4.1 Maturity Model Level	**Process Capability Based on ISO/IEC 15504**	**Context**
5 Optimised—Processes have been refined to a level of good practice, based on the results of continuous improvement and maturity modelling with other enterprises. IT is used in an integrated way to automate the workflow, providing tools to improve quality and effectiveness, making the enterprise quick to adapt.	**Level 5: Optimising process**—The level 4 predictable process is continuously improved to meet relevant current and projected business goals.	Enterprise View—Corporate Knowledge
4 Managed and measurable—Management monitors and measures compliance with procedures and takes action where processes appear not to be working effectively. Processes are under constant improvement and provide good practice. Automation and tools are used in a limited or fragmented way.	**Level 4: Predictable process**—The level 3 established process now operates within defined limits to achieve its process outcomes.	
3 Defined process—Procedures have been standardised and documented, and communicated through training. It is mandated that these processes should be followed; however, it is unlikely that deviations will be detected. The procedures themselves are not sophisticated, but are the formalisation of existing practices.	**Level 3: Established process**—The level 2 managed process is now implemented using a defined process that is capable of achieving its process outcomes.	
	Level 2: Managed process—The level 1 performed process is now implemented in a managed fashion (planned, monitored and adjusted) and its work products are appropriately established, controlled and maintained.	Instance View—Individual Knowledge
2 Repeatable but intuitive—Processes have developed to the stage where similar procedures are followed by different people undertaking the same task. There is no formal training or communication of standard procedures, and responsibility is left to the individual. There is a high degree of reliance on the knowledge of individuals and, therefore, errors are likely.	**Level 1: Performed process**—The implemented process achieves its process purpose. **Remark: It is possible that some classified as Maturity Model 1 will be classified as 15504 0, if the process outcomes are not achieved.**	
1 Initial/Ad hoc—There is evidence that the enterprise has recognised that the issues exist and need to be addressed. There are, however, no standardised processes; instead, there are ad hoc approaches that tend to be applied on an individual or case-by-case basis. The overall approach to management is disorganised.		
0 Non-existent—Complete lack of any recognisable processes. The enterprise has not even recognised that there is an issue to be addressed.	**Level 0: Incomplete process**—The process is not implemented or fails to achieve its purpose.	

Figure 21—Comparison Table of Maturity Attributes (COBIT 4.1) and Process Attributes (COBIT 5)									
	COBIT 5 Process Capability Attribute								
COBIT 4.1 Maturity Attribute	Process Performance	Performance Management	Work Product Management	Process Definition	Process Deployment	Process Management	Process Control	Process Innovation	Process Optimisation
Awareness and communication				■	■	■			
Policies, plans and procedures							■		
Tools and automation				■	■				
Skills and expertise				■	■				
Responsibility and accountability				■					
Goals setting and measurement		■				■			

Benefits of the Changes

The benefits of the COBIT 5 process capability model, compared to the COBIT 4.1 maturity models, include:
• Improved focus on the process being performed, to confirm that it is actually achieving its purpose and delivering its required outcomes as expected.

- Simplified content through elimination of duplication, because the COBIT 4.1 maturity model assessment required the use of a number of specific components, including the generic maturity model, process maturity models, control objectives and process controls to support process assessment.
- Improved reliability and repeatability of process capability assessment activities and evaluations, reducing debates and disagreements between stakeholders on assessment results.
- Increased usability of process capability assessment results, because the new model establishes a basis for more formal, rigorous assessments to be performed, for both internal and potential external purposes.
- Compliance with a generally accepted process assessment standard and therefore strong support for the process assessment approach in the market.

Performing Process Capability Assessments in COBIT 5

The ISO/IEC 15504 standard specifies that process capability assessments can be performed for various purposes and with varying degrees of rigour. Purposes can be internal, with a focus on comparisons between enterprise areas and/or process improvement for internal benefit, or they can be external, with a focus on formal assessment, reporting and certification.

The COBIT 5 ISO/IEC 15504-based assessment approach continues to facilitate the following objectives that have been a key COBIT approach since 2000 to:
- Enable the governance body and management to benchmark process capability.
- Enable high-level 'as-is' and 'to-be' health checks to support the governance body and management investment decision making with regard to process improvement.
- Provide gap analysis and improvement planning information to support definition of justifiable improvement projects.
- Provide the governance body and management with assessment ratings to measure and monitor current capabilities.

This section describes how a high-level assessment can be performed with the COBIT 5 process capability model to achieve these objectives.

The assessment distinguishes between assessing capability level 1 and the higher levels. Indeed, as described previously, process capability level 1 describes whether a process achieves its intended purpose, and is therefore a very important level to achieve—as well as foundational in enabling higher capability levels to be reached.

Assessing whether the process achieves its goals—or, in other words, achieves capability level 1—can be done by:
1. Reviewing the process outcomes as they are described for each process in the detailed process descriptions, and using the ISO/IEC 15504 rating scale to assign a rating to what degree each objective is achieved. This scale consists of the following ratings:
   - **N** (Not achieved)—There is little or no evidence of achievement of the defined attribute in the assessed process. (0 to 15 percent achievement)
   - **P** (Partially achieved)—There is some evidence of an approach to, and some achievement of, the defined attribute in the assessed process. Some aspects of achievement of the attribute may be unpredictable. (15 to 50 percent achievement)
   - **L** (Largely achieved)—There is evidence of a systematic approach to, and significant achievement of, the defined attribute in the assessed process. Some weakness related to this attribute may exist in the assessed process. (50 to 85 percent achievement)
   - **F** (Fully achieved)—There is evidence of a complete and systematic approach to, and full achievement of, the defined attribute in the assessed process. No significant weaknesses related to this attribute exist in the assessed process. (85 to 100 percent achievement)
2. In addition, the process (governance or management) practices can be assessed using the same rating scale, expressing the extent to which the base practices are applied.
3. To further refine the assessment, the work products also may be taken into consideration to determine the extent to which a specific assessment attribute has been achieved.

Although defining target capability levels is up to each enterprise to decide, many enterprises will have the ambition to have all their processes achieve capability level 1. (Otherwise, what would be the point of having these processes?) If this level is not achieved, the reasons for not achieving this level are immediately obvious from the approach explained above, and an improvement plan can be defined:
1. If a required process outcome is not consistently achieved, the process does not meet its objective and needs to be improved.
2. The assessment of the process practices will reveal which practices are lacking or failing, enabling implementation and/or improvement of those practices to take place and allowing all process outcomes to be achieved.

For higher process capability levels, the generic practices are used, taken from ISO/IEC 15504:2. They provide generic descriptions for each of the capability levels.

Page intentionally left blank

Page intentionally left blank

APPENDIX A
REFERENCES

The following frameworks, standards and other guidance were used as reference material and input for the development of COBIT 5.

Association for Project Management (APM); *APM Introduction to Programme Management*, Latimer, Trend and Co., UK, 2007

British Standards Institute (BSI), BS25999:2007 Business Continuity Management Standard, UK, 2007

CIO Council, *Federal Enterprise Architecture* (FEA), ver 1.0, USA, 2005

European Commission, *The Commission Enterprise IT Architecture Framework (CEAF)*, Belgium, 2006

Kotter, John; *Leading Change*, Harvard Business School Press, USA, 1996

HM Government, Best Management Practice Portfolio, *Managing Successful Programmes (MSP)*, UK, 2009

HM Government, Best Management Practice Portfolio, *PRINCE2®*, UK, 2009

HM Government, Best Management Practice Portfolio, *Information Technology Infrastructure Library (ITIL®)*, 2011

International Organization for Standardization (ISO), 9001:2008 Quality Management Standard, Switzerland, 2008

ISO/International Electrotechnical Commission (IEC), 20000:2006 IT Service Management Standard, Switzerland, 2006

ISO/IEC, 27005:2008, Information Security Risk Management Standard, Switzerland, 2008

ISO/IEC, 38500:2008, Corporate Governance of Information Technology Standard, Switzerland, 2008

King Code of Governance Principles (King III), South Africa, 2009

Organisation for Economic Co-operation and Development (OECD), *OECD Principles of Corporate Governance*, France, 2004

The Open Group, TOGAF® 9, UK, 2009

Project Management Institute, Project Management Body of Knowledge (PMBOK2®), USA, 2008

UK Financial Reporting Council, 'Combined Code on Corporate Governance', UK, 2009

Page intentionally left blank

Page intentionally left blank

APPENDIX B
DETAILED MAPPING ENTERPRISE GOALS—IT-RELATED GOALS

The COBIT 5 goals cascade is explained in chapter 2.

The purpose of the mapping table in **figure 22** is to demonstrate how enterprise goals are supported (or translate into) IT-related goals. For that reason, the table contains the following information:
• In the columns, all 17 generic enterprise goals defined in COBIT 5, grouped by BSC dimension
• In the rows, all 17 IT-related goals, also grouped in IT BSC dimensions
• A mapping of how each enterprise goal is supported by IT-related goals. This mapping is expressed using the following scale:
  – 'P' stands for primary, when there is an important relationship, i.e., the IT-related goal is a primary support for the enterprise goal.
  – 'S' stands for secondary, when there is still a strong, but less important, relationship, i.e., the IT-related goal is a secondary support for the enterprise goal.

EXAMPLE 7—MAPPING TABLE

The mapping table suggests that one would normally expect that:
• Enterprise goal 7. Business service continuity and availability will:
  – Primarily depend on the achievement of the IT-related goals:
    • 04 Managed IT-related business risk
    • 10 Security of information, processing infrastructure and applications
    • 14 Availability of reliable and useful information for decision making
  – Also depend, but to a lesser degree, on the achievement of the IT-related goals:
    • 01 Alignment of IT and business strategy
    • 07 Delivery of IT services in line with business requirements
    • 08 Adequate use of applications, information and technology solutions
• Using the table in the opposite direction, achieving the IT-related goal 09. IT agility will contribute to the achievement of several enterprise goals:
  – Primarily, the enterprise goals:
    • 2. Portfolio of competitive products and services
    • 8. Agile responses to a changing business environment
    • 11. Optimisation of business process functionality
    • 17. Product and business innovation culture
  – To a lesser degree, the enterprise goals:
    • 1. Stakeholder value of business investments
    • 3. Managed business risk (safeguarding of assets)
    • 6. Customer-oriented service culture
    • 13. Managed business change programmes
    • 14. Operational and staff productivity
    • 16. Skilled and motivated people

The table was created based on the following inputs:
• Research by the University of Antwerp Management School IT Alignment and Governance Research Institute
• Additional reviews and expert opinions obtained during the development and review process of COBIT 5

When using the table in figure 22, please consider the remarks made in chapter 2 on how to use the COBIT 5 goals cascade.

			Enterprise Goal																
			Stakeholder value of business investments	Portfolio of competitive products and services	Managed business risk (safeguarding of assets)	Compliance with external laws and regulations	Financial transparency	Customer-oriented service culture	Business service continuity and availability	Agile responses to a changing business environment	Information-based strategic decision making	Optimisation of service delivery costs	Optimisation of business process functionality	Optimisation of business process costs	Managed business change programmes	Operational and staff productivity	Compliance with internal policies	Skilled and motivated people	Product and business innovation culture
			1.	2.	3.	4.	5.	6.	7.	8.	9.	10.	11.	12.	13.	14.	15.	16.	17.
		IT-related Goal	Financial					Customer					Internal					Learning and Growth	
Financial	01	Alignment of IT and business strategy	P	P	S			P	S	P	P	S	P	S	P			S	S
	02	IT compliance and support for business compliance with external laws and regulations			S	P											P		
	03	Commitment of executive management for making IT-related decisions	P	S	S					S	S		S		P			S	S
	04	Managed IT-related business risk			P	S			P	S		S			S		S	S	
	05	Realised benefits from IT-enabled investments and services portfolio	P	P				S		S		S	S	P		S			S
	06	Transparency of IT costs, benefits and risk	S		S		P				S	P		P					
Customer	07	Delivery of IT services in line with business requirements	P	P	S	S		P	S	P	S		P	S	S			S	S
	08	Adequate use of applications, information and technology solutions	S	S	S			S	S		S	S	P	S		P		S	S
Internal	09	IT agility	S	P	S			S		P			P		S	S		S	P
	10	Security of information, processing infrastructure and applications			P	P			P								P		
	11	Optimisation of IT assets, resources and capabilities	P	S						S			P	S	P	S	S		S
	12	Enablement and support of business processes by integrating applications and technology into business processes	S	P	S			S		S			S	P	S	S	S		S
	13	Delivery of programmes delivering benefits, on time, on budget, and meeting requirements and quality standards	P	S	S			S					S		S	P			
	14	Availability of reliable and useful information for decision making	S	S	S	S			P		P		S						
	15	IT compliance with internal policies			S	S											P		
Learning and Growth	16	Competent and motivated business and IT personnel	S	S	P			S		S						P		P	S
	17	Knowledge, expertise and initiatives for business innovation	S	P				S		P	S		S		S			S	P

Figure 22—Mapping COBIT 5 Enterprise Goals to IT-related Goals

APPENDIX C
DETAILED MAPPING IT-RELATED GOALS—IT-RELATED PROCESSES

This appendix contains the mapping table between the IT-related goals and how these are supported by IT-related processes, as part of the goals cascade explained in chapter 2.

Figure 23 contains:
• In the columns, all 17 generic IT-related goals defined in chapter 2, grouped in IT BSC dimensions
• In the rows, all 37 COBIT 5 processes, grouped by domain
• A mapping of how each IT-related goal is supported by a COBIT 5 IT-related process. This mapping is expressed using the following scale:
  – 'P' stands for primary, when there is an important relationship, i.e., the COBIT 5 process is a primary support for the achievement of an IT-related goal.
  – 'S' stands for secondary, when there is still a strong, but less important, relationship, i.e., the COBIT 5 process is a secondary support for the IT-related goal.

EXAMPLE 8—AP013 MANAGE SECURITY
The process AP013 *Manage security* will contribute: • Primarily, to the achievement of the IT-related goals: – 02 IT compliance and support for business compliance with external laws and regulations – 04 Managed IT-related business risk – 06 Transparency of IT costs, benefits and risk – 10 Security of information, processing infrastructure and applications – 14 Availability of reliable and useful information for decision making • To a lesser degree, to the achievement of the IT-related goals: – 07 Delivery of IT services in line with business requirements – 08 Adequate use of applications, information and technology solutions

The table was created based on the following inputs:
• Research by the University of Antwerp Management School IT Alignment and Governance Research Institute
• Additional reviews and expert opinions obtained during the development and review process of COBIT 5

When using the table in figure 23, please consider the remarks made in chapter 2 on how to use the COBIT 5 goals cascade.

Figure 23—Mapping COBIT 5 IT-related Goals to Processes

IT-related Goal

	COBIT 5 Process	01 Alignment of IT and business strategy	02 IT compliance and support for business compliance with external laws and regulations	03 Commitment of executive management for making IT-related decisions	04 Managed IT-related business risk	05 Realised benefits from IT-enabled investments and services portfolio	06 Transparency of IT costs, benefits and risk	07 Delivery of IT services in line with business requirements	08 Adequate use of applications, information and technology solutions	09 IT agility	10 Security of information, processing infrastructure and applications	11 Optimisation of IT assets, resources and capabilities	12 Enablement and support of business processes by integrating applications and technology into business processes	13 Delivery of programmes delivering benefits, on time, on budget, and meeting requirements and quality standards	14 Availability of reliable and useful information for decision making	15 IT compliance with internal policies	16 Competent and motivated business and IT personnel	17 Knowledge, expertise and initiatives for business innovation
		Financial						Customer		Internal							Learning and Growth	
Evaluate, Direct and Monitor	EDM01 Ensure Governance Framework Setting and Maintenance	P	S	P	S	S	S	P		S	S	S	S	S	S	S	S	S
	EDM02 Ensure Benefits Delivery	P		S		P	P	P	S			S	S	S			S	P
	EDM03 Ensure Risk Optimisation	S	S	S	P		P	S	S		P			S	S	P	S	S
	EDM04 Ensure Resource Optimisation	S		S	S	S	S	S	S	P		P		S			P	S
	EDM05 Ensure Stakeholder Transparency	S	S	P			P	P						S	S	S		S
Align, Plan and Organise	APO01 Manage the IT Management Framework	P	P	S	S			S		P	S	P	S	S	S	S	P	P
	APO02 Manage Strategy	P		S	S	S		P	S	S		S	S		S	S	S	P
	APO03 Manage Enterprise Architecture	P		S	S	S	S	S	S	P	S	P	S		S			S
	APO04 Manage Innovation	S				P		P	P	P		P	S		S			P
	APO05 Manage Portfolio	P		S	S	P	S	S	S	S		S		P				S
	APO06 Manage Budget and Costs	S		S	S	P	P	S	S			S		S				
	APO07 Manage Human Resources	P	S	S	S			S		S	S	P		P		S	P	P
	APO08 Manage Relationships	P		S	S	S	S	P	S			S	P	S			S	P
	APO09 Manage Service Agreements	S			S	S	S	P	S	S	S	S			S	P	S	
	APO10 Manage Suppliers		S		P	S	S	P	S	P	S	S			S	S	S	S
	APO11 Manage Quality	S	S		S	P		P	S	S		S		P	S	S	S	S
	APO12 Manage Risk		P		P		P	S	S	S	P			P	S	S	S	S
	APO13 Manage Security		P		P		P	S	S		P					P		

Figure 23—Mapping COBIT 5 IT-related Goals to Processes (cont.)

IT-related Goal legend:
01 — Alignment of IT and business strategy
02 — IT compliance and support for business compliance with external laws and regulations
03 — Commitment of executive management for making IT-related decisions
04 — Managed IT-related business risk
05 — Realised benefits from IT-enabled investments and services portfolio
06 — Transparency of IT costs, benefits and risk
07 — Delivery of IT services in line with business requirements
08 — Adequate use of applications, information and technology solutions
09 — IT agility
10 — Security of information, processing infrastructure and applications
11 — Optimisation of IT assets, resources and capabilities
12 — Enablement and support of business processes by integrating applications and technology into business processes
13 — Delivery of programmes delivering benefits, on time, on budget, and meeting requirements and quality standards
14 — Availability of reliable and useful information for decision making
15 — IT compliance with internal policies
16 — Competent and motivated business and IT personnel
17 — Knowledge, expertise and initiatives for business innovation

Column groups: 01–06 Financial · 07–08 Customer · 09–15 Internal · 16–17 Learning and Growth

COBIT 5 Process	01	02	03	04	05	06	07	08	09	10	11	12	13	14	15	16	17
Build, Acquire and Implement																	
BAI01 Manage Programmes and Projects	P		S	P	P	S	S	S			S		P			S	S
BAI02 Manage Requirements Definition	P	S	S	S	S		P	S	S	S	S	P	S	S			S
BAI03 Manage Solutions Identification and Build	S			S	S		P	S			S	S	S	S			S
BAI04 Manage Availability and Capacity				S	S		P	S	S		P			S	P		S
BAI05 Manage Organisational Change Enablement	S		S		S		S	P	S		S	S	P				P
BAI06 Manage Changes			S	P	S		P	S	S	P	S	S	S	S	S		S
BAI07 Manage Change Acceptance and Transitioning				S	S		S	P	S			P	S	S	S		S
BAI08 Manage Knowledge	S				S		S	S	P	S	S			S		S	P
BAI09 Manage Assets		S		S		P	S			S	S	P		S	S		
BAI10 Manage Configuration		P		S		S		S	S	S	P			P	S		
Deliver, Service and Support																	
DSS01 Manage Operations		S		P	S		P	S	S	S	P			S	S	S	S
DSS02 Manage Service Requests and Incidents				P			P	S		S				S	S		S
DSS03 Manage Problems		S		P	S		P	S	S		P	S		P	S		S
DSS04 Manage Continuity	S	S		P	S		P	S	S	S	S	S		P	S	S	S
DSS05 Manage Security Services	S	P		P			S	S		P	S	S		S	S		
DSS06 Manage Business Process Controls		S		P			P	S		S	S	S	S	S	S	S	S
Monitor, Evaluate and Assess																	
MEA01 Monitor, Evaluate and Assess Performance and Conformance	S	S	S	P	S	S	P	S	S	S	P			S	S	P	S
MEA02 Monitor, Evaluate and Assess the System of Internal Control		P		P			S	S	S		S				S	P	S
MEA03 Monitor, Evaluate and Assess Compliance With External Requirements		P		P	S			S			S				S		S

Page intentionally left blank

Page intentionally left blank

APPENDIX D
STAKEHOLDER NEEDS AND ENTERPRISE GOALS

Chapter 4 showed the individual steps of the goals cascade, starting from stakeholder needs down to enabler goals. Chapter 2 included a table with typical governance and management questions on IT. From a stakeholder point of view it is interesting to know how these questions relate to the enterprise goals. For that reason, **figure 24** is included; it shows how a list of internal stakeholder needs can be linked to the enterprise goals.

This table can be used to help setting and prioritising specific enterprise goals or IT-related goals, based on specific stakeholder needs. The same precautions should be used when using these tables as with the other goals cascade tables, i.e., every enterprise's individual situation differs, and these tables should not be used in a mechanical way, but only as a suggested generic set of relationships. In **figure 24**, the intersection of a stakeholder need and enterprise goal is filled in if that need should be considered for that goal.

Figure 24—Mapping COBIT 5 Enterprise Goals to Governance and Management Questions

STAKEHOLDER NEEDS	1. Stakeholder value of business investments	2. Portfolio of competitive products and services	3. Managed business risk (safeguarding of assets)	4. Compliance with external laws and regulations	5. Financial transparency	6. Customer-oriented service culture	7. Business service continuity and availability	8. Agile responses to a changing business environment	9. Information-based strategic decision making	10. Optimisation of service delivery costs	11. Optimisation of business process functionality	12. Optimisation of business process costs	13. Managed business change programmes	14. Operational and staff productivity	15. Compliance with internal policies	16. Skilled and motivated people	17. Product and business innovation culture
How do I get value from the use of IT? Are end users satisfied with the quality of the IT service?	■	■				■	■						■			■	■
How do I manage performance of IT?		■		■				■	■	■	■		■				
How can I best exploit new technology for new strategic opportunities?	■							■	■				■			■	
How do I best build and structure my IT department?								■		■	■		■		■	■	
How dependent am I on external providers? How well are IT outsourcing agreements being managed? How do I obtain assurance over external providers?			■	■						■							
What are the (control) requirements for information?				■					■						■		
Did I address all IT-related risk?			■	■											■		
Am I running an efficient and resilient IT operation?				■			■										
How do I control the cost of IT? How do I use IT resources in the most effective and efficient manner? What are the most effective and efficient sourcing options?										■		■		■			
Do I have enough people for IT? How do I develop and maintain their skills, and how do I manage their performance?										■	■	■				■	
How do I get assurance over IT?				■												■	

Figure 24—Mapping COBIT 5 Enterprise Goals to Governance and Management Questions (cont.)

STAKEHOLDER NEEDS	Stakeholder value of business investments	Portfolio of competitive products and services	Managed business risk (safeguarding of assets)	Compliance with external laws and regulations	Financial transparency	Customer-oriented service culture	Business service continuity and availability	Agile responses to a changing business environment	Information-based strategic decision making	Optimisation of service delivery costs	Optimisation of business process functionality	Optimisation of business process costs	Managed business change programmes	Operational and staff productivity	Compliance with internal policies	Skilled and motivated people	Product and business innovation culture
	1.	2.	3.	4.	5.	6.	7.	8.	9.	10.	11.	12.	13.	14.	15.	16.	17.
Is the information I am processing well secured?				■			■								■		
How do I improve business agility through a more flexible IT environment?	■															■	■
Do IT projects fail to deliver what they promised—and if so, why? Is IT standing in the way of executing the business strategy?	■	■	■					■			■	■					
How critical is IT to sustaining the enterprise? What do I do if IT is not available?	■		■				■										
What concrete vital primary business processes are dependent on IT, and what are the requirements of business processes?	■	■									■						
What has been the average overrun of the IT operational budgets? How often and how much do IT projects go over budget?					■					■				■			
How much of the IT effort goes to fighting fires rather than to enabling business improvements?		■	■									■					
Are sufficient IT resources and infrastructure available to meet required enterprise strategic objectives?		■			■					■							
How long does it take to make major IT decisions?	■	■			■			■									
Are the total IT effort and investments transparent?		■		■	■										■		
Does IT support the enterprise in complying with regulations and service levels? How do I know whether I am compliant with all applicable regulations?				■											■		

APPENDIX E
MAPPING OF COBIT 5 WITH THE MOST RELEVANT RELATED STANDARDS AND FRAMEWORKS

Introduction

This appendix compares COBIT 5 to the most relevant and used standards and frameworks in the governance space. For ISO/IEC 38500 this is done through a comparison based on the ISO/IEC 38500 principles; for the other comparisons a table format is used in which the COBIT 5 processes are mapped against the equivalent contents in the referred standard or framework.

COBIT 5 and ISO/IEC 38500

The following summarises how COBIT 5 supports adoption of the standard's principles and implementation approach. The standard, *ISO/IEC 38500:2008–Corporate governance of information technology,* is based on six key principles. The practical implications of each principle are explained here, together with how COBIT 5 guidance enables good practice.

ISO/IEC 38500 Principles
PRINCIPLE 1—RESPONSIBILITY
What this means in practice:
The business (customer) and IT (provider) should collaborate in a partnership model utilising effective communications based on a positive and trusted relationship and demonstrating clarity regarding responsibility and accountability. For larger enterprises, an IT executive committee (also referred to as the IT strategy committee) acting on behalf of the board and chaired by a board member is a very effective mechanism for evaluating, directing and monitoring the use of IT in the enterprise and for advising the board on critical IT issues. Directors of small and medium-sized enterprises with a simpler command structure and shorter communication paths need to take a more direct approach when overseeing IT activities. In all cases, appropriate governance organisational structures, roles and responsibilities are required to be mandated from the governing body, providing clear ownership and accountability for important decisions and tasks. This should include relationships with key third-party IT service providers.

How ISACA's guidance enables good practice:
1. The COBIT 5 framework defines a number of enablers for governance of enterprise IT. The 'process' enabler and the 'organisational structures' enabler, combined with the RACI[13] charts, are particularly relevant in this context. They strongly advocate assignment of responsibilities, and provide example roles and responsibilities for board members and management for all key related processes and activities.
2. *COBIT 5 Implementation* explains the responsibilities of stakeholders and other involved parties when implementing or enhancing IT governance arrangements.
3. COBIT 5 has two levels of monitoring. The first level is relevant in a governance context. The process EDM05 *Ensure stakeholder transparency* explains the director's role in monitoring and evaluating IT governance and IT performance with a generic method for establishing goals and objectives and related metrics.

PRINCIPLE 2—STRATEGY
What this means in practice:
IT strategic planning is a complex and critical undertaking requiring close co-ordination amongst enterprisewide business unit and IT strategic plans. It is also vital to prioritise the plans most likely to achieve the desired benefits and to allocate resources effectively. High-level goals need to be translated into achievable tactical plans, ensuring minimal failures and surprises. The goal is to deliver value in support of strategic objectives while considering the associated risk in relation to the board's risk appetite. While it is important to cascade plans in a top-down fashion, the plans must also be flexible and adaptable to meet rapidly changing business requirements and IT opportunities.

Furthermore, the presence or absence of IT capabilities can either enable or hinder business strategies; therefore, IT strategic planning should include transparent and appropriate planning of IT capabilities. This should include assessment of the ability of the current IT infrastructure and human resources to support future business requirements and consideration of future technological developments that might enable competitive advantage and/or optimise costs. IT resources include relationships with many external product vendors and service providers, some of whom likely play a critical role in supporting the business. Governance of strategic sourcing is thus a very significant strategic planning activity requiring executive-level direction and oversight.

---

[13] RACI charts outline who is Responsible, Accountable, Consulted and Informed for a task.

How ISACA's guidance enables good practice:
1. COBIT 5 provides specific guidance on managing IT investments and (specifically, in the process EDM02 *Ensure benefits delivery* in the governance domain) how strategic objectives should be supported by appropriate business cases.
2. The COBIT 5 APO domain explains the processes required for the effective planning and organisation of internal and external IT resources, including strategic planning, technology and architecture planning, organisational planning, innovation planning, portfolio management, investment management, risk management, relationship management and quality management. The alignment of business and IT goals is also explained, with generic examples showing how they support strategic objectives for all IT-related processes based on industrywide research.
3. The exercise of identifying and aligning enterprise goals and IT-related goals presents a better understanding of the cascading relationship amongst enterprise goals, IT-related goals and enablers, which include IT processes. It presents a solid and strong list of 17 generic enterprise goals and 17 generic IT-related goals, validated and prioritised amongst different sectors. Together with the linking information between both, it provides a good basis on which to build a generic cascade from business goals to IT goals.

PRINCIPLE 3—ACQUISITION
What this means in practice:
IT solutions exist to support business processes and therefore care must be taken to not consider IT solutions in isolation or as just a 'technology' project or service. On the other hand, an inappropriate choice of technology architecture, a failure to maintain a current and appropriate technical infrastructure, or an absence of skilled human resources can result in project failure, an inability to sustain business operations or a reduction in value to the business. Acquisitions of IT resources should be considered as a part of wider IT-enabled business change. The acquired technology must also support and operate with existing and planned business processes and IT infrastructures. Implementation is also not just a technology issue, but rather a combination of organisational change, revised business processes, training and enabling the change. Therefore, IT projects should be undertaken as part of wider enterprisewide change programmes that include other projects satisfying the full range of activities required to help ensure a successful outcome.

How ISACA's guidance enables good practice:
1. The COBIT 5 EDM domain provides guidance on governing and managing IT-enabled business investments through their complete life cycle (acquisition, implementation, operation and decommissioning). The APO05 *Manage portfolio* process addresses how to apply effective portfolio and programme management of such investments to help ensure that benefits are realised and costs are optimised.
2. The COBIT 5 APO domain provides guidance for planning for acquisition, including investment planning, risk management, programme and project planning, and quality planning.
3. The COBIT 5 BAI domain provides guidance on the processes required to acquire and implement IT solutions, covering defining requirements, identifying feasible solutions, preparing documentation, and training and enabling users and operations to run the new systems. In addition, guidance is provided to help ensure that the solutions are tested and controlled properly as the change is applied to the operational business and IT environment.
4. The COBIT 5 MEA domain and process EDM05 include guidance on how directors can monitor and evaluate the acquisition process, and internal controls to help ensure that acquisitions are properly managed and executed.

PRINCIPLE 4—PERFORMANCE
What this means in practice:
Effective performance measurement depends on two key aspects being addressed: the clear definition of performance goals and the establishment of effective metrics to monitor achievement of goals. A performance measurement process is also required to help ensure that performance is monitored consistently and reliably. Effective governance is achieved when goals are set from the top down and aligned with high-level, approved business goals, and metrics are established from the bottom up and aligned in a way that enables the achievement of goals at all levels to be monitored by each layer of management. Two critical governance success factors are the approval of goals by stakeholders, and the acceptance of accountability for achievement of goals by directors and managers. IT is a complex and technical topic; therefore, it is important to achieve transparency by expressing goals, metrics and performance reports in language meaningful to the stakeholders so that appropriate actions can be taken.

How ISACA's guidance enables good practice:
1. The COBIT 5 framework provides generic examples of goals and metrics for the full range of IT-related processes and the other enablers, and shows how they relate to business goals, enabling enterprises to adapt them for their own specific use.
2. COBIT 5 provides management with guidance on setting IT objectives in alignment with business goals and describes how to monitor performance of these objectives using goals and metrics. Process capability can be assessed using an ISO/IEC 15504 compliance capability assessment model.

3. Two key COBIT 5 processes provide specific guidance:
  – APO02 *Manage strategy* focusses on setting goals.
  – APO09 *Manage service agreements* focusses on defining appropriate services and service goals and documenting them in service level agreements.
4. In process MEA01 *Monitor, evaluate and assess performance and conformance*, COBIT 5 provides guidance on responsibilities of executive management for this activity.
5. The planned *COBIT 5 for Assurance* guide will explain how assurance professionals can provide independent assurance to directors regarding IT performance.

PRINCIPLE 5—CONFORMANCE
What this means in practice:
In today's global marketplace, enabled by the Internet and advanced technologies, enterprises need to comply with a growing number of legal and regulatory requirements. Because of corporate scandals and financial failures in recent years, there is a heightened awareness in the boardroom of the existence and implications of tougher laws and regulations. Stakeholders require increased assurance that enterprises are complying with laws and regulations and conforming to good corporate governance practice in their operating environment. In addition, because IT has enabled seamless business processes between enterprises, there is also a growing need to help ensure that contracts include important IT-related requirements in areas such as privacy, confidentiality, intellectual property and security.

Directors need to ensure that compliance with external requirements is dealt with as a part of strategic planning rather than as a costly afterthought. They also need to set the tone at the top and establish policies and procedures for their management and staff to follow, to ensure that the goals of the enterprise are realised, risk is minimised and compliance is achieved. Top management must strike an appropriate balance between performance and conformance, ensuring that performance goals do not jeopardise compliance and, conversely, that the conformance regime is appropriate and does not overly restrict the operation of the business.

How ISACA's guidance enables good practice:
1. The COBIT 5 governance and management practices provide a basis for establishing an appropriate control environment in the enterprise. The process capability assessments enable management to evaluate and benchmark IT process capability.
2. COBIT 5 process APO02 *Manage strategy* helps ensure that there is alignment between the IT plan and the overall business objectives, including governance requirements.
3. COBIT 5 process MEA02 *Monitor, evaluate and assess the system of internal control* enables directors to assess whether controls are adequate to meet compliance requirements.
4. COBIT 5 process MEA03 *Monitor, evaluate and assess compliance with external requirements* helps ensure that external compliance requirements are identified, directors set the direction for compliance, and IT compliance itself is monitored, assessed and reported as a part of overall conformance to enterprise requirements.
5. The planned *COBIT 5 for Assurance* guide explains how auditors can provide independent assurance of compliance and adherence to internal policies derived from internal directives or external legal, regulatory or contractual requirements, confirming that any corrective actions to address any compliance gaps have been taken by the responsible process owner in a timely manner.

PRINCIPLE 6—HUMAN BEHAVIOUR
What this means in practice:
The implementation of any IT-enabled change, including IT governance itself, usually requires significant cultural and behavioural change within enterprises as well as with customers and business partners. This can create fear and misunderstanding amongst staff, so implementation needs to be managed carefully if personnel are to remain positively engaged. Directors must clearly communicate goals and be seen as positively supporting the proposed changes. Training and skills enhancement of personnel are key aspects of change—especially given the rapidly moving nature of technology. People are affected by IT at all levels in an enterprise, as stakeholders, managers and users, or as specialists providing IT-related services and solutions to the business. Beyond the enterprise, IT affects customers and business partners and increasingly enables self-service and automated intercompany transactions within countries and across borders. While IT-enabled business processes bring new benefits and opportunities, they also carry increasing types of risk. Issues such as privacy and fraud are growing concerns for individuals, and these and other types of risk need to be managed if people are to trust the IT systems they use. Information systems can also dramatically affect working practices by automating manual procedures.

How ISACA's guidance enables good practice:
The following COBIT 5 enablers (which include processes) provide guidance on requirements relating to human behaviour:
1. COBIT 5 enablers include people, skills and competencies, and culture, ethics and behaviour. For each enabler a model is presented on how to deal with the enabler, illustrated by examples.

2. COBIT 5 process APO07 *Manage human resources* explains how the performance of individuals should be aligned with corporate goals, how IT specialist skills should be maintained, and how roles and responsibilities should be defined.
3. COBIT 5 process BAI02 *Manage requirements definition* helps ensure design of applications to meet human operation and use requirements.
4. COBIT 5 processes BAI05 *Manage organisational change enablement* and BAI08 *Manage knowledge* help ensure that users are enabled to use systems effectively.

In addition, ISACA provides four certifications for professionals performing key roles related to the governance of IT, and for which the body of knowledge is substantially covered by the COBIT 5 contents:
• Certified in the Governance of Enterprise IT® (CGEIT®)
• Certified Information Systems Auditor® (CISA®)
• Certified Information Security Manager® (CISM®)
• Certified in Risk and Information Systems Control™ (CRISC™)

Holders of these certifications have demonstrated both capability and experience in performing such roles.

ISO/IEC 38500 Evaluate, Direct and Monitor
HOW ISACA'S GUIDANCE ENABLES GOOD PRACTICE:
The governance domain in the COBIT 5 process model has five processes, and each of these processes has EDM practices defined. This is the main location in COBIT 5 where governance-related activities are defined.

Comparison With Other Standards

COBIT 5 was developed taking into account a number of other standards and frameworks; these standards are listed in appendix A.

COBIT 5: Enabling Processes contains a high-level mapping between each COBIT 5 process and the most relevant parts of related standards and frameworks containing additional guidance.

In this section a brief discussion of each framework or standard is included, indicating to which areas and domains in COBIT 5 it relates.

ITIL® V3 2011 and ISO/IEC 20000
The following COBIT 5 areas and domains are covered by ITIL V3 2011 and ISO/IEC 20000:
• A subset of processes in the DSS domain
• A subset of processes in the BAI domain
• Some processes in the APO domain

ISO/IEC 27000 Series
The following COBIT 5 areas and domains are covered by ISO/IEC 27000:
• Security- and risk-related processes in the EDM, APO and DSS domains
• Various security-related activities within processes in other domains
• Monitor and evaluating activities from the MEA domain

ISO/IEC 31000 Series
The following COBIT 5 areas and domains are covered by ISO/IEC 31000:
• Risk-management-related processes in the EDM and APO domains

TOGAF®
The following COBIT 5 areas and domains are covered by TOGAF:
• Resource-related processes in the EDM (governance) domain—The TOGAF components of an Architecture Board, Architecture Governance and Architecture Maturity Models map to resource optimisation.
• The enterprise architecture process in the APO domain. At the core of TOGAF is the Architecture Development Method (ADM) cycle, which maps to the COBIT 5 practices of developing an architecture vision (ADM Phase A), defining reference architectures (ADM Phases B, C, D), selecting opportunities and solutions (ADM Phase E), and defining architecture implementation (ADM Phases F, G). A number of TOGAF components map to the COBIT 5 practice of providing enterprise architecture services. These include:
  – ADM Requirements Management
  – Architecture Principles

– Stakeholder Management
– Business Transformation Readiness Assessment
– Risk Management
– Capability-based Planning
– Architecture Compliance
– Architecture Contracts

Capability Maturity Model Integration (CMMI) (development)

The following COBIT 5 areas and domains are covered by CMMI:
• Application–building- and acquisition-related processes in the BAI domain
• Some organisational and quality-related processes from the APO domain

PRINCE2®

The following COBIT 5 areas and domains are covered by PRINCE2:
• Portfolio-related processes in the APO domain
• Programme and project management processes in the BAI domain

Figure 25 depicts the relative coverage between COBIT 5 and the other standards and frameworks.

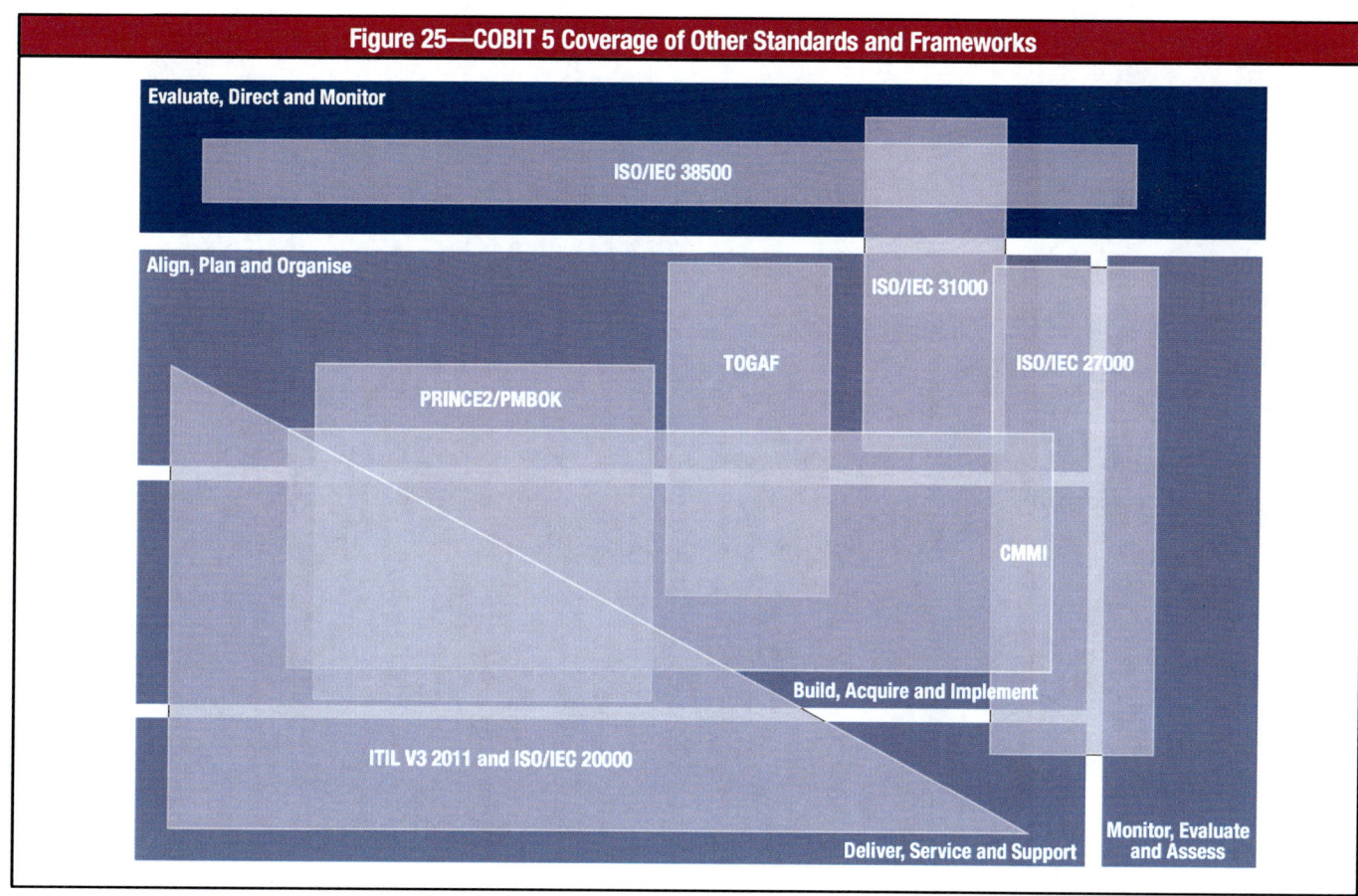

Figure 25—COBIT 5 Coverage of Other Standards and Frameworks

Page intentionally left blank

Page intentionally left blank

APPENDIX F
COMPARISON BETWEEN THE COBIT 5 INFORMATION MODEL AND COBIT 4.1 INFORMATION CRITERIA

How do the seven information criteria of COBIT 4.1—effectiveness, efficiency, integrity, reliability, availability, confidentiality, compliance—relate to the information quality categories and dimensions of the COBIT 5 information enablers, as shown in appendix G, **figure 32**?

The following table contains two columns:
• The first column lists each of the seven COBIT 4.1 information criteria.
• The second column lists the COBIT 5 alternatives, i.e., the corresponding information enabler goal(s).

Figure 26—COBIT 5 Equivalents to COBIT 4.1 Information Criteria	
COBIT 4.1 Information Criteria	**COBIT 5 Equivalent**
Effectiveness	Information is effective if it meets the needs of the information consumer who uses the information for a specific task. If the information consumer can perform the task with the information, then the information is effective. This corresponds to the following information quality goals: appropriate amount, relevance, understandability, interpretability, objectivity.
Efficiency	Whereas effectiveness considers the information as a product, efficiency relates more to the process of obtaining and using information, so it aligns to the 'information as a service' view. If information that meets the needs of the information consumer is obtained and used in an easy way (i.e., it takes few resources—physical effort, cognitive effort, time, money), then the use of information is efficient. This corresponds to the following information quality goals: believability, accessibility, ease of operation, reputation.
Integrity	If information has integrity, then it is free of error and complete. It corresponds to the following information quality goals: completeness, accuracy.
Reliability	Reliability is often seen as a synonym of accuracy; however, it can also be said that information is reliable if it is regarded as true and credible. Compared to integrity, reliability is more subjective, more related to perception, and not just factual. It corresponds to the following information quality goals: believability, reputation, objectivity.
Availability	Availability is one of the information quality goals under the accessibility and security heading.
Confidentiality	Confidentiality corresponds to the restricted access information quality goal.
Compliance	Compliance in the sense that information must conform to specifications is covered by any of the information quality goals, depending on the requirements. Compliance to regulations is most often a goal or requirement of the use of the information, not so much an inherent quality of information.

This table shows that all information criteria from COBIT 4.1 are covered by COBIT 5; however, the COBIT 5 information model allows definition of an additional set of criteria, hence adding value to the COBIT 4.1 criteria.

Page intentionally left blank

APPENDIX G
DETAILED DESCRIPTION OF COBIT 5 ENABLERS

Introduction

This section contains a more detailed discussion of the seven categories of enablers that are part of the COBIT 5 framework, which were initially described in chapter 5, and are repeated in **figure 27**.

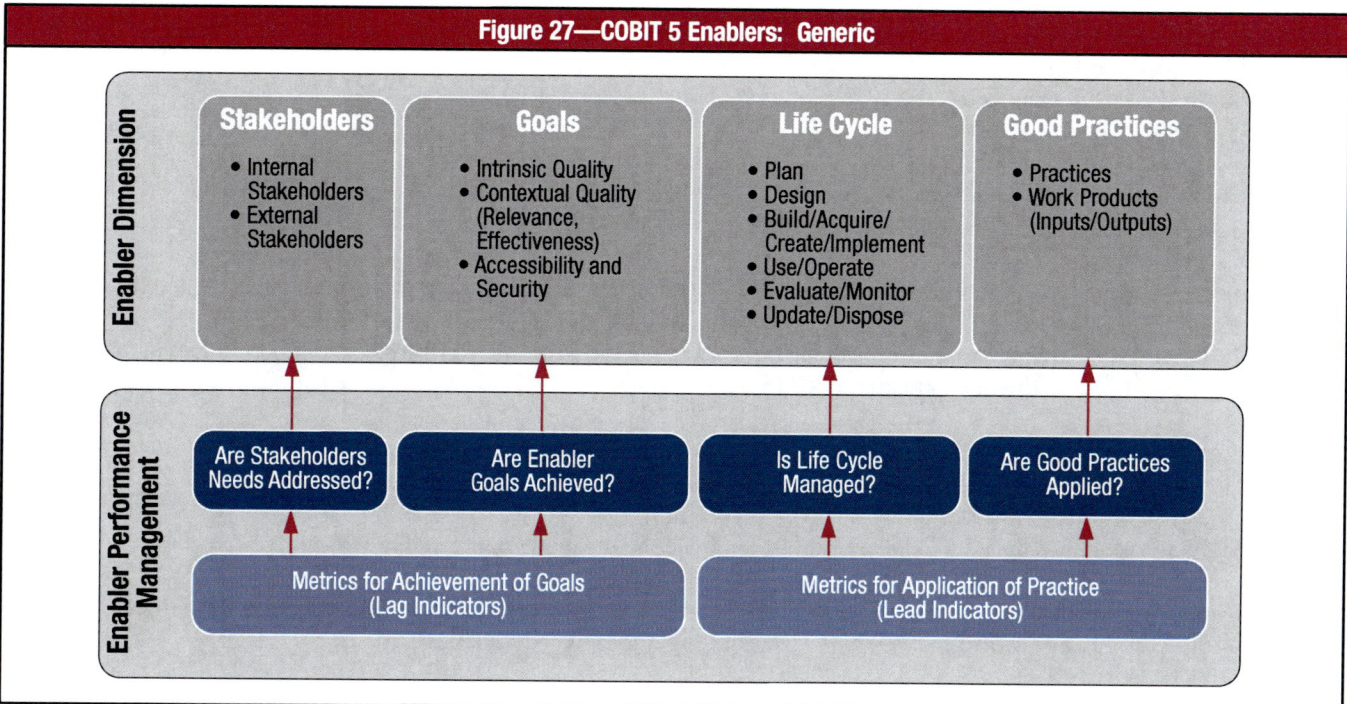

Figure 27—COBIT 5 Enablers: Generic

Enabler Dimensions
The four common dimensions for enablers are:
• **Stakeholders**—Each enabler has stakeholders, i.e., parties who play an active role and/or have an interest in the enabler. For example, processes have different parties who execute process activities and/or who have an interest in the process outcomes; organisational structures have stakeholders—each with his/her own roles and interests—that are part of the structures. Stakeholders can be internal or external to the enterprise, all having their own, sometimes conflicting, interests and needs. Stakeholders' needs translate to enterprise goals, which in turn translate to IT-related goals for the enterprise. A list of stakeholders is shown in **figure 7**.
• **Goals**—Each enabler has a number of goals, and enablers provide value by the achievement of these goals. Goals can be defined in terms of:
  – Expected outcomes of the enabler
  – Application or operation of the enabler itself

The enabler goals are the final step in the COBIT 5 goals cascade. Goals can be further split up in different categories such as:
  – **Intrinsic quality**—The extent to which enablers work accurately, objectively and provide accurate, objective and reputable results
  – **Contextual quality**—The extent to which enablers and their outcomes are fit for the purpose, given the context in which they operate. For example, outcomes should be relevant, complete, current, appropriate, consistent, understandable and easy to use.
  – **Access and security**—The extent to which enablers and their outcomes are accessible and secured:
    • Enablers are available when, and if, needed.
    • Outcomes are secured, i.e., access is restricted to those entitled to and needing it.
• **Life cycle**—Each enabler has a life cycle, from inception through an operational/useful life until disposal. This applies to information, structures, processes, policies, etc. The phases of the life cycle consist of:
  – Plan (which includes concepts development and concepts selection)
  – Design
  – Build/acquire/create/implement
  – Use/operate

– Evaluate/monitor
– Update/dispose
• **Good practices**—For each of the enablers, good practices can be defined. Good practices support the achievement of the enabler goals. Good practices provide examples or suggestions on how best to implement the enabler, and what work products or inputs and outputs are required. COBIT 5 provides examples of good practices for some enablers in COBIT 5 (e.g., processes). For other enablers, guidance from other standards, frameworks, etc., can be used.

Enabler Performance Management

Enterprises expect positive outcomes from the application and use of enablers. To manage performance of the enablers, the following questions have to be monitored and subsequently answered—based on metrics—on a regular basis:
• Are stakeholder needs addressed?
• Are enabler goals achieved?
• Is the enabler life cycle managed?
• Are good practices applied?

The first two bullets deal with the actual outcome of the enabler, and the metrics used to measure the extent to which the goals are achieved can be called 'lag indicators'.

The last two bullets deal with the actual functioning of the enabler itself, and metrics for this can be called 'lead indicators'.

For each enabler there is a separate section, which starts with a drawing similar to **figure 27**, but including a number of specific elements for the enabler at hand, indicated in red and **bold**.

Next, each of the four components is discussed in more detail, discussing specific components and relationships with other enablers.

A number of examples have been included as well to illustrate the meaning and use of the enablers.

The purpose of this section is to give more insight in the COBIT 5 framework, and how the enabler concept can be applied to implement and improve governance and management over enterprise IT.

COBIT 5 Enabler: Principles, Policies and Frameworks

Principles and policies refer to the communication mechanisms put in place to convey the governing body's and management's direction and instructions. The specifics for the principles, policies, and frameworks enabler compared to the generic enabler description are shown in **figure 28**.

The principles, policies and frameworks model shows:
• **Stakeholders**—Stakeholders for principles and policies can be internal and external to the enterprise. They include the board and executive management, compliance officers, risk managers, internal and external auditors, service providers and customers, and regulatory agencies. The stakes are twofold: Some stakeholders define and set policies, others have to align to, and comply with, policies.
• **Goals and metrics**—Principles, policies and frameworks are instruments to communicate the rules of the enterprise, in support of the governance objectives and enterprise values, as defined by the board and executive management. Principles need to be:
  – Limited in number
  – Put in simple language, expressing as clearly as possible the core values of the enterprise

Policies provide more detailed guidance on how to put principles into practice and they influence how decision making aligns with the principles. Good policies are:
  – Effective—They achieve the stated purpose.
  – Efficient—They ensure that principles are implemented in the most efficient way.
  – Non-intrusive—They appear logical for those who have to comply with them, i.e., they do not create unnecessary resistance.

Access to policies—Is there a mechanism in place that provides easy access to policies for all stakeholders? In other words, do stakeholders know where to find policies?

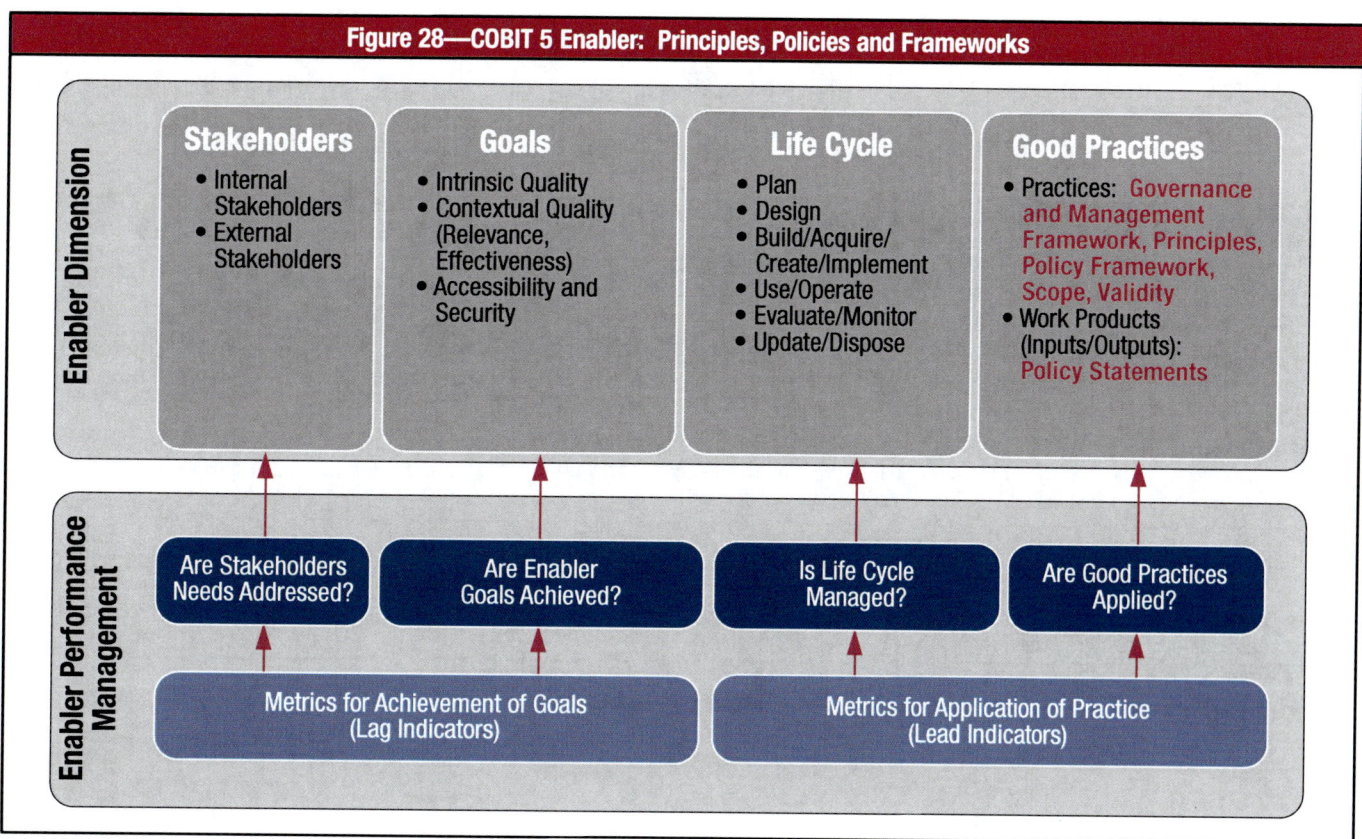

Figure 28—COBIT 5 Enabler: Principles, Policies and Frameworks

Governance and management frameworks should provide management with structure, guidance, tools, etc., that allow proper governance and management of enterprise IT. The frameworks should be:
  – Comprehensive, covering all required areas
  – Open and flexible, allowing adaptation to the enterprise's specific situation
  – Current, i.e., reflecting the current direction of the enterprise and the current governance objectives
  – Available and accessible to all stakeholders

• **Life cycle**—Policies have a life cycle that has to support the achievement of the defined goals. Frameworks are key because they provide a structure to define consistent guidance. For example, a policy framework provides the structure in which a consistent set of policies can be created and maintained, and it also provides an easy point of navigation within and between individual policies.

Depending on the external environment in which the enterprise operates, there can be varying degrees of regulatory requirements for strong internal control and, as a consequence, a strong policy framework. A key attention point to be taken into account regarding frameworks and policies is the currency of policies—If and when policies are reviewed and updated, are there strong mechanisms in place to ensure that people are aware of these updates, that the newest version is easily accessible (see previous point), and that obsolete information is properly archived or disposed?

• **Good practices:**
  – Good practice requires that policies be part of an overall governance and management framework, providing a (hierarchical) structure into which all policies should fit and clearly make the link to the underlying principles.
  – As part of the policy framework, the following items need to be described:
    • Scope and validity
    • The consequences of failing to comply with the policy
    • The means for handling exceptions
    • The manner in which compliance with the policy will be checked and measured
  – Generally recognised governance and management frameworks can provide valuable guidance on the actual statements to be included in policies.
  – Policies should be aligned with the enterprise's risk appetite. Policies are a key component of an enterprise's system of internal control, whose purpose it is to manage and contain risk. As part of risk governance activities, the enterprise's risk appetite is defined, and this risk appetite should be reflected in the policies. A risk-averse enterprise has stricter policies than a risk-aggressive enterprise.
  – Policies need to be revalidated and/or updated at regular intervals.

• **Relationships with other enablers**—The links with other enablers include:
  – Principles, policies and frameworks should reflect the culture and ethical values of the enterprise, and they should encourage the desired behaviour; hence, there is a strong link with the culture, ethics and behaviour enabler.
  – Process practices and activities are the most important vehicle for executing policies.
  – Organisational structures can define and implement policies within their span of control, and their activities are also defined by policies.
  – Policies are also information, so all good practices applying to information apply to policies as well.

EXAMPLE 9—SOCIAL MEDIA

An enterprise is considering how to deal with the fast-rising use of social media and pressure from its staff to have full access. Until now, the organisation has been conservative or restrictive in granting access to this kind of service, mainly for security reasons.

There is pressure from different sides to take another position with regard to social media. Staff members want similar levels of access as they have from home, and the organisation itself also wants to use and exploit the benefits of social media for marketing and public awareness-raising purposes.

The decision is taken to define a policy on the use of social media on the enterprise's networks and systems, including laptops provided by the enterprise to its staff members. The new policy fits in the existing policy framework under the category of 'acceptable use policies', and it is more relaxed than previous policies. As a consequence, communication is developed to explain the reasons for the new policy. At the same time, there is also an impact on some other enablers:

• Staff members need to learn how to deal with the new media to avoid creating embarrassing situations for the enterprise. They need to learn the appropriate behaviour in line with the new direction that the enterprise is taking and develop the appropriate skills.

• A number of processes with regard to security need to be changed. Access is opened up to these media, so security settings and configurations have to change and possibly some compensating measures need to be defined.

Note: COBIT 5 is an example of a framework as described in this enabler.

COBIT 5 Enabler: Processes

The specifics for the processes enabler compared to the generic enabler description are shown in **figure 29.**

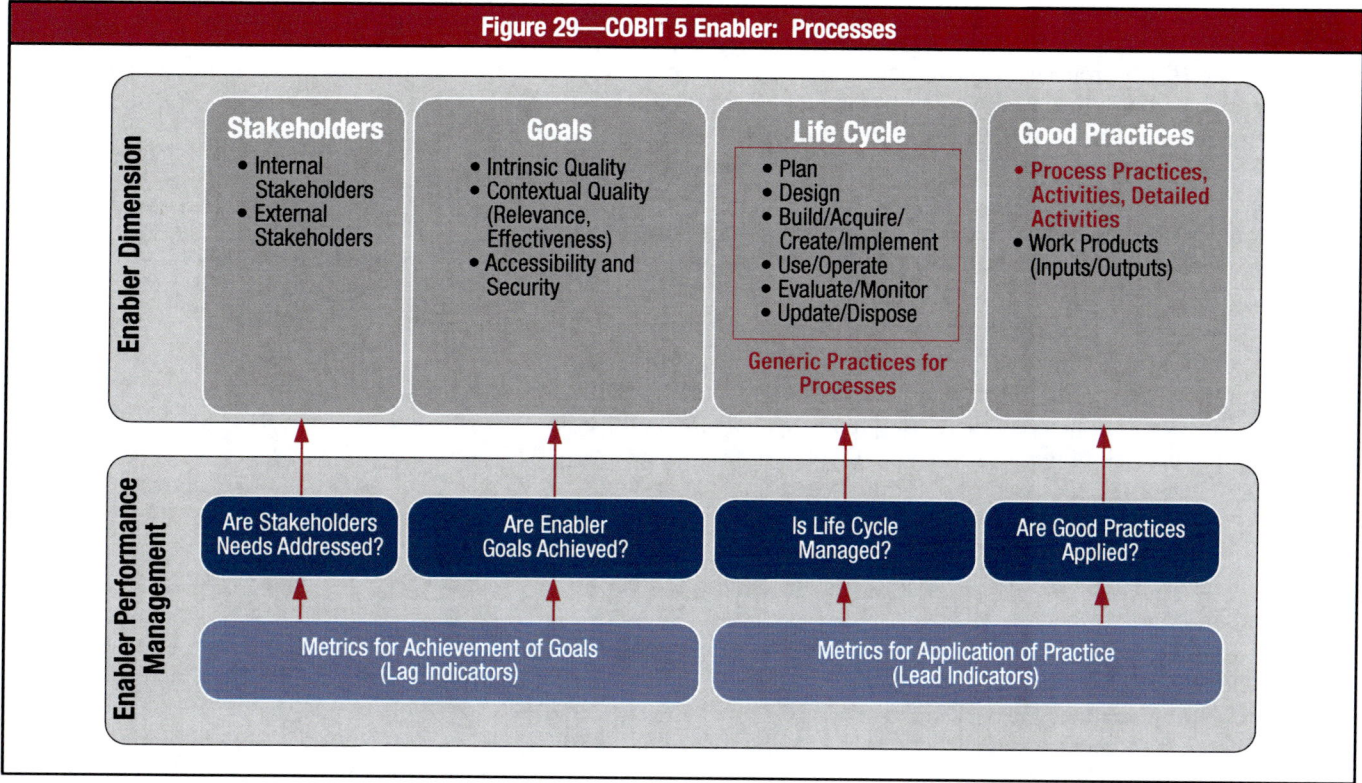

Figure 29—COBIT 5 Enabler: Processes

A process is defined as **'a collection of practices influenced by the enterprise's policies and procedures that takes inputs from a number of sources (including other processes), manipulates the inputs and produces outputs (e.g., products, services)'.**

The processes model shows:
- **Stakeholders**—Processes have internal and external stakeholders, with their own roles; stakeholders and their responsibility levels are documented in RACI charts. External stakeholders include customers, business partners, shareholders and regulators. Internal stakeholders include the board, management, staff and volunteers.
- **Goals**—Process goals are defined as 'a statement describing the desired outcome of a process. An outcome can be an artefact, a significant change of a state or a significant capability improvement of other processes'. They are part of the goals cascade, i.e., process goals support IT-related goals, which in turn support enterprise goals.

Process goals can be categorised as:
- **Intrinsic goals**—Does the process have intrinsic quality? Is it accurate and in line with good practice? Is it compliant with internal and external rules?
- **Contextual goals**—Is the process customised and adapted to the enterprise's specific situation? Is the process relevant, understandable, easy to apply?
- **Accessibility and security goals**—The process remains confidential, when required, and is known and accessible to those who need it.

At each level of the goals cascade, hence also for processes, metrics are defined to measure the extent to which goals are achieved. Metrics can be defined as 'a quantifiable entity that allows the measurement of the achievement of a process goal. Metrics should be SMART—specific, measurable, actionable, relevant and timely'.

To manage the enabler effectively and efficiently, metrics need to be defined to measure the extent to which the expected outcomes are achieved. In addition, a second aspect of performance management of the enabler describes the extent to which good practice is applied. Here also, associated metrics can be defined to help with the management of the enabler.

- **Life cycle**—Each process has a life cycle. It is defined, created, operated, monitored, and adjusted/updated or retired. Generic process practices such as those defined in the COBIT process assessment model based on ISO/IEC 15504 can assist with defining, running, monitoring and optimising processes.
- **Good practices**—*COBIT 5: Enabling Processes* contains a process reference model, in which process internal good practices are described in growing levels of detail: practices, activities and detailed activities:[14]

Practices:
- For each COBIT 5 process, the governance/management practices provide a complete set of high-level requirements for effective and practical governance and management of enterprise IT. They are:
  - Statements of actions to deliver benefits, optimise the level of risk and optimise the use of resources
  - Aligned with relevant generally accepted standards and good practices
  - Generic and therefore needing to be adapted for each enterprise
  - Covering business and IT role players in the process (end-to-end)
- The enterprise governance body and management need to make choices relative to these governance and management practices by:
  - Selecting those that are applicable and deciding on those that will be implemented
  - Adding and/or adapting practices where required
  - Defining and adding non-IT-related practices for integration in business processes
  - Choosing how to implement them (frequency, span, automation, etc.)
  - Accepting the risk of not implementing those that may apply

Activities—In COBIT, the main actions taken to operate the process
- They are defined as 'guidance to achieve management practices for successful governance and management of enterprise IT'. The COBIT 5 activities provide the how, why and what to implement for each governance or management practice to improve IT performance and/or address IT solution and service delivery risk. This material is of use to:
  - Management, service providers, end users and IT professionals who need to plan, build, run or monitor enterprise IT
  - Assurance professionals who may be asked for their opinions regarding current or proposed implementations or necessary improvements
- A complete set of generic and specific activities that provide one approach consisting of all the steps that are necessary and sufficient for achieving the key governance practice (GP)/management practice (MP). They provide high-level guidance, at a level below the GP/MP, for assessing actual performance and for considering potential improvements. The activities:
  – Describe a set of necessary and sufficient action-oriented implementation steps to achieve a GP/MP
  – Consider the inputs and outputs of the process
  – Are based on generally accepted standards and good practices
  – Support establishment of clear roles and responsibilities
  – Are non-prescriptive, and need to be adapted and developed into specific procedures appropriate for the enterprise

Detailed activities—The activities may not be at a sufficient level of detail for implementation, and further guidance may need to be:
  – Obtained from specific relevant standards and good practices such as ITIL, the ISO/IEC 27000 series and PRINCE2
  – Developed as more detailed or specific activities as additional developments in the COBIT 5 product family itself

Inputs and outputs—The COBIT 5 inputs and outputs are the process work products/artefacts considered necessary to support operation of the process. They enable key decisions, provide a record and audit trail of process activities, and enable follow-up in the event of an incident. They are defined at the key governance/management practice level, may include some work products used only within the process, and are often essential inputs to other processes.[15]

External good practices can exist in any form or level of detail, and mostly refer to other standards and frameworks. Users can refer to these external good practices at all times, knowing that COBIT is aligned with these standards where relevant, and mapping information will be made available.

Enabler Performance Management

Enterprises expect positive outcomes from the application and use of enablers. To manage performance of the enablers, the following questions will have to be monitored and answered—based on metrics—on a regular basis:
- Are stakeholder needs addressed?
- Are enabler goals achieved?
- Is the enabler life cycle managed?
- Are good practices applied?

---

[14] Only practices and activities are developed under the current project. The more detailed levels are subject to additional development(s), e.g., the various professional guides may provide more detailed guidance for their areas. Also, further guidance can be obtained through related standards and frameworks, as indicated in the detailed process descriptions.

[15] The illustrative COBIT 5 inputs and outputs should not be regarded as an exhaustive list because additional information flows could be defined, depending on a particular enterprise's environment and process framework.

In the case of the process enabler, the first two bullets deal with the actual outcome of the process. The metrics used to measure the extent to which the goals are achieved can be called 'lag indicators'. In *COBIT 5: Enabling Processes*, a number of metrics are defined per process goal.

The last two bullets deal with the actual functioning of the enabler itself, and metrics for this can be called 'lead indicators'.

Process capability level—COBIT 5 includes an ISO/IEC 15504-based process capability assessment scheme. This is discussed in chapter 8 of COBIT 5 and further guidance is available from separate ISACA COBIT 5 publications. In brief, the process capability level measures both achievement of goals and application of good practice.

Relationships with other enablers—Links between processes and the other enabler categories exist through the following relationships:
• Processes need information (as one of the types of inputs) and can produce information (as a work product).
• Processes need organisational structures and roles to operate, as expressed through the RACI charts, e.g., IT steering committee, enterprise risk committee, board, audit, CIO, CEO.
• Processes produce, and also require, service capabilities (infrastructure, applications, etc.).
• Processes can, and will, depend on other processes.
• Processes produce, or need, policies and procedures to ensure consistent implementation and execution.
• Cultural and behavioural aspects determine how well processes are executed.

Example of Process Enabler in Practice
Example 10 illustrates the process enabler, its interconnections and the enabler dimensions. The example builds on example 7 earlier in the document.

COBIT 5 Process Reference Model
GOVERNANCE AND MANAGEMENT PROCESSES
One of the guiding principles in COBIT is the distinction made between governance and management. In line with this principle, every enterprise would be expected to implement a number of governance processes and a number of management processes to provide comprehensive governance and management of enterprise IT.

When considering processes for governance and management in the context of the enterprise, the difference between types of processes lies within the objectives of the processes:
• **Governance processes**—Governance processes deal with the stakeholder governance objectives—value delivery, risk optimisation and resource optimisation—and include practices and activities aimed at evaluating strategic options, providing direction to IT and monitoring the outcome (EDM—in line with the ISO/IEC 38500 standard concepts).
• **Management processes**—In line with the definition of management, practices and activities in management processes cover the responsibility areas of PBRM enterprise IT, and they have to provide end-to-end coverage of IT.

EXAMPLE 10—PROCESS ENABLER INTERCONNECTIONS

An organisation has appointed 'process managers' for IT-related processes, charged with defining and operating effective and efficient IT-related processes, in the context of good governance and management of enterprise IT.

Initially, the process managers will focus on the process enabler, considering the enabler dimensions:
- **Stakeholders:** Process stakeholders include all process actors, i.e., all parties that are responsible, accountable, consulted or informed (RACI) for, or during, process activities. For this, a RACI chart as described in *COBIT 5: Enabling Processes* can be used.
- **Goals:** For each process, adequate goals and related metrics need to be defined. For example, for process APO08 *Manage relationships* (in *COBIT 5: Enabling Processes*) one can find a set of process goals and metrics such as:
  - **Goal:** Business strategies, plans and requirements are well understood, documented and approved.
    - **Metric:** Percent of programmes aligned with enterprise business requirements/priorities
  - **Goal:** Good relationships exist between the enterprise and the IT department.
    - **Metric:** Ratings of user and IT personnel satisfaction surveys
- **Life cycle:** Each process has a life cycle, i.e., it has to be created, executed and monitored, and adjusted when required. Eventually, processes cease to exist. In this case, the process managers would need to design and define the process first. They can use several elements from *COBIT 5: Enabling Processes* to design processes, i.e., to define responsibilities and to break the process down into practices and activities, and define process work products (inputs and outputs). In a later stage, the process needs to be made more robust and efficient, and for that purpose the process managers can raise the capability level of the process. The ISO/IEC 15504-inspired COBIT 5 Process Capability Model and the process capability attributes can be used for that purpose such as:
  - Process capability level 2 requires the achievement of two attributes: Performance Management and Work Product Management. The first attribute requires a number of planning-phase-related activities:
    - Objectives for the performance of the process are defined.
    - Performance of the process is planned.
    - Responsibilities for performing the process are defined.
    - Resources are identified.
    - Etc.

    The same capability level prescribes a number of activities for the 'monitoring' phase of the process life cycle such as:
    - Performance of the process is monitored.
    - Performance of the process is adjusted to meet plans.
    - Etc.
  - The same approach can be used to obtain guidance for the different phases in the life cycle from the different performance capability attributes at increasing levels of capability of processes.
- **Good practice:** COBIT 5 describes in ample detail good practice for processes in *COBIT 5: Enabling Processes*, as mentioned under the previous point. Inspiration and example processes can be found there, covering the full spectrum of required activities for good governance and management of enterprise IT.

In addition to the guidance on the process enabler, the process managers can decide to look at a number of other enablers such as:
- The RACI charts describe roles and responsibilities. Other enablers allow drilling down on this dimension such as:
  - In the skills and competencies enabler, the required skills and competencies for each role can be defined and appropriate goals (e.g., technical and behavioural skill levels) and associated metrics can be defined.
  - The RACI chart also contains a number of organisational structures. These structures can be further elaborated in the organisational structures enabler, where a more detailed description of the structure can be provided, expected outcomes and related metrics can be defined (e.g., decisions), good practices can be defined (e.g., span of control, operating principles of the structure, level of authority).
- Principles and policies will formalise the processes and prescribe why the process exists, to whom it is applicable, and how the process is to be used. This is the focus area of the principles and policies enabler.

Although the outcome of both types of processes is different and intended for a different audience, internally, from the context of the process itself, all processes require 'planning', 'building or implementation', 'execution' and 'monitoring' activities within the process.

COBIT 5 PROCESS REFERENCE MODEL
COBIT 5 is not prescriptive, but from the previous text it is clear that it advocates that enterprises implement governance and management processes such that the key areas are covered, as shown in **figure 30**.

In theory, an enterprise can organise its processes as it sees fit, as long as the basic governance and management objectives are covered. Smaller enterprises may have fewer processes; larger and more complex enterprises may have many processes, all to cover the same objectives.

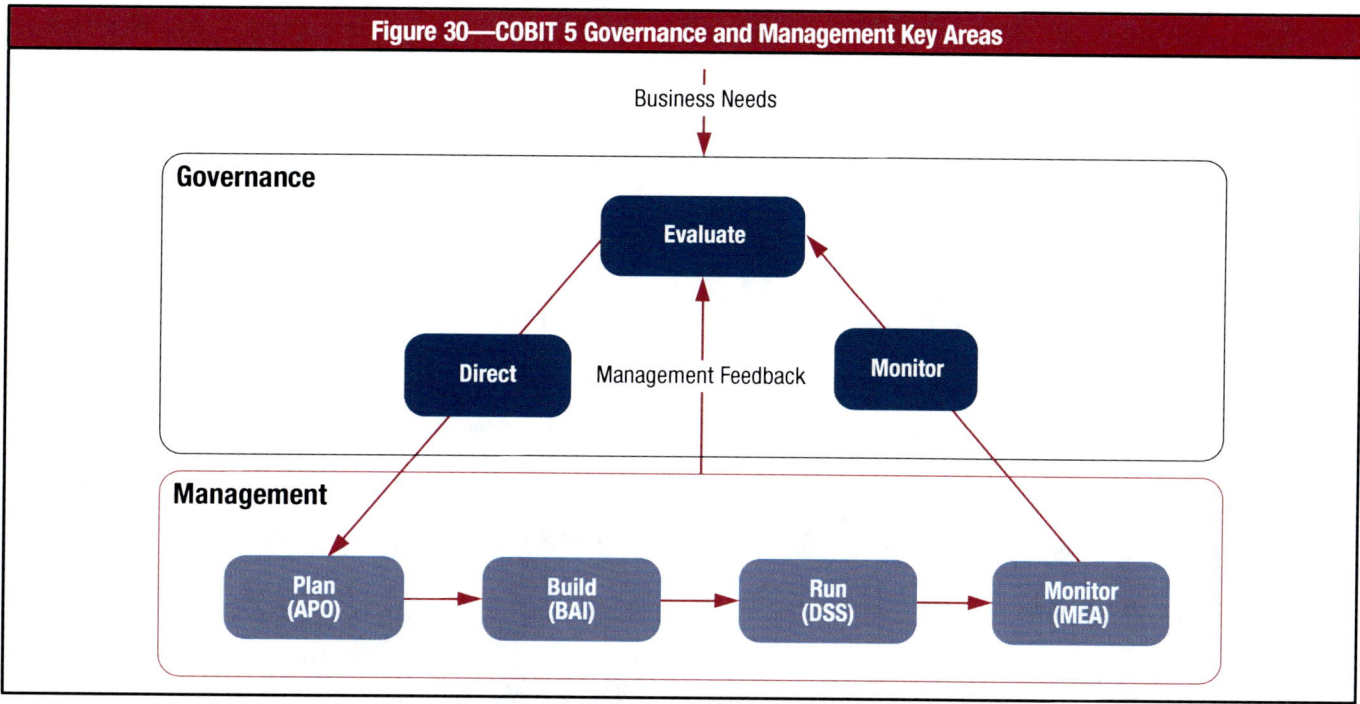

Figure 30—COBIT 5 Governance and Management Key Areas

Notwithstanding the previous text, COBIT 5 includes a process reference model, defining and describing in detail a number of governance and management processes. It provides a process reference model that represents all of the processes normally found in an enterprise relating to IT activities, offering a common reference model understandable to operational IT and business managers. The proposed process model is a complete, comprehensive model, but it is not the only possible process model. Each enterprise must define its own process set, taking into account the specific situation.

Incorporating an operational model and a common language for all parts of the enterprise involved in IT activities is one of the most important and critical steps towards good governance. It also provides a framework for measuring and monitoring IT performance, communicating with service providers, and integrating best management practices.

The COBIT 5 process reference model subdivides the governance and management processes of enterprise IT into two main areas of activity—governance and management—divided into domains of processes:
• **Governance**—This domain contains five governance processes; within each process, EDM practices are defined.
• **Management**—These four domains are in line with the responsibility areas of PBRM (an evolution of the COBIT 4.1 domains), and they provide end-to-end coverage of IT. Each domain contains a number of processes, as in COBIT 4.1 and previous versions. Although, as described previously, most of the processes require 'planning', 'implementation', 'execution' and 'monitoring' activities within the process or within the specific issue being addressed (e.g., quality, security), they are placed in domains in line with what is generally the most relevant area of activity when regarding IT at the enterprise level.

In COBIT 5, the processes also cover the full scope of business and IT activities related to the governance and management of enterprise IT, thus making the process model truly enterprisewide.

The COBIT 5 process reference model is the successor to the COBIT 4.1 process model, with the Risk IT and Val IT process models integrated as well. **Figure 31** shows the complete set of 37 governance and management processes within COBIT 5. The details of all processes, according to the process model described previously, are included in *COBIT 5: Enabling Processes*.

Figure 31—COBIT 5 Process Reference Model

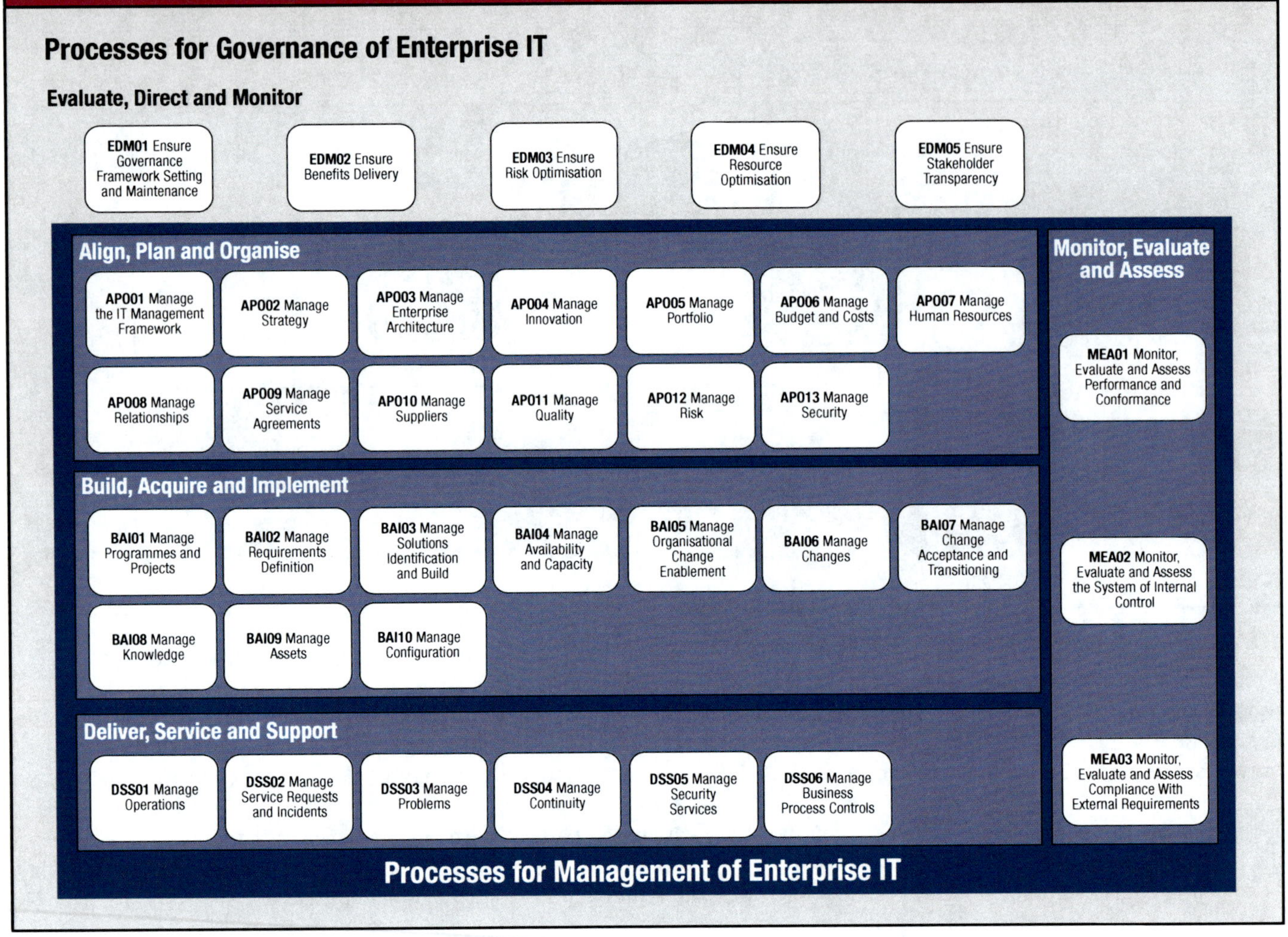

Processes for Governance of Enterprise IT

Evaluate, Direct and Monitor

- **EDM01** Ensure Governance Framework Setting and Maintenance
- **EDM02** Ensure Benefits Delivery
- **EDM03** Ensure Risk Optimisation
- **EDM04** Ensure Resource Optimisation
- **EDM05** Ensure Stakeholder Transparency

Align, Plan and Organise

- **APO01** Manage the IT Management Framework
- **APO02** Manage Strategy
- **APO03** Manage Enterprise Architecture
- **APO04** Manage Innovation
- **APO05** Manage Portfolio
- **APO06** Manage Budget and Costs
- **APO07** Manage Human Resources
- **APO08** Manage Relationships
- **APO09** Manage Service Agreements
- **APO10** Manage Suppliers
- **APO11** Manage Quality
- **APO12** Manage Risk
- **APO13** Manage Security

Build, Acquire and Implement

- **BAI01** Manage Programmes and Projects
- **BAI02** Manage Requirements Definition
- **BAI03** Manage Solutions Identification and Build
- **BAI04** Manage Availability and Capacity
- **BAI05** Manage Organisational Change Enablement
- **BAI06** Manage Changes
- **BAI07** Manage Change Acceptance and Transitioning
- **BAI08** Manage Knowledge
- **BAI09** Manage Assets
- **BAI10** Manage Configuration

Deliver, Service and Support

- **DSS01** Manage Operations
- **DSS02** Manage Service Requests and Incidents
- **DSS03** Manage Problems
- **DSS04** Manage Continuity
- **DSS05** Manage Security Services
- **DSS06** Manage Business Process Controls

Monitor, Evaluate and Assess

- **MEA01** Monitor, Evaluate and Assess Performance and Conformance
- **MEA02** Monitor, Evaluate and Assess the System of Internal Control
- **MEA03** Monitor, Evaluate and Assess Compliance With External Requirements

Processes for Management of Enterprise IT

COBIT 5 Enabler: Organisational Structures

The specifics for the organisational structures enabler compared to the generic enabler description are shown in **figure 32**.

The organisational structures model shows:
- **Stakeholders**—Organisational structures stakeholders can be internal and external to the enterprise, and they include the individual members of the structure, other structures, organisational entities, clients, suppliers and regulators. Their roles vary, and include decision making, influencing and advising. The stakes of each of the stakeholders also vary, i.e., what interest do they have in the decisions made by the structure?
- **Goals**—The goals for the organisational structures enabler itself would include having a proper mandate, well-defined operating principles and application of other good practices. The outcome of the organisational structures enabler should include a number of good activities and decisions.
- **Life cycle**—An organisational structure has a life cycle. It is created, exists and is adjusted, and finally it can be disbanded. During its inception, a mandate—a reason and purpose for its existence—has to be defined.
- **Good practices**—A number of good practices for organisational structures can be distinguished such as:
  - Operating principles—The practical arrangements regarding how the structure will operate, such as frequency of meetings, documentation and housekeeping rules
  - Composition—Structures have members, who are internal or external stakeholders.
  - Span of control—The boundaries of the organisational structure's decision rights
  - Level of authority/decision rights—The decisions that the structure is authorised to take
  - Delegation of authority—The structure can delegate (a subset of) its decision rights to other structures reporting to it.
  - Escalation procedures—The escalation path for a structure describes the required actions in case of problems in making decisions.

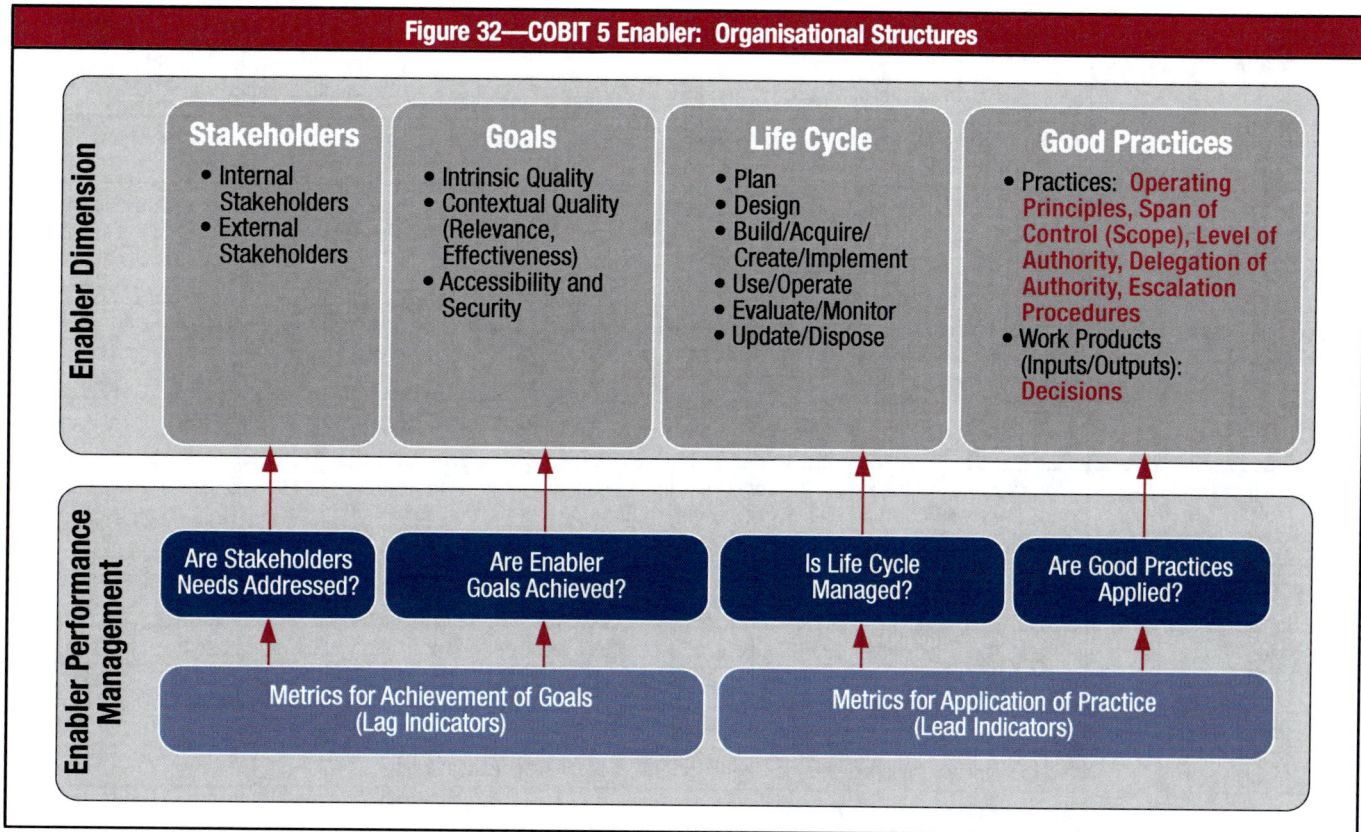

Figure 32—COBIT 5 Enabler: Organisational Structures

Relationships with other enablers—The links with other enablers include:
- RACI charts link process activities to organisational structures and/or individual roles in the enterprise. They describe the level of involvement of each role for each process practice: responsible, accountable, consulted or informed.
- Culture, ethics and behaviour determine the efficiency and effectiveness of organisational structures and their decisions.
- Composition of organisational structures should take into account and require the appropriate skill set of its members.
- The mandate and operating principles of organisational structures are guided by the policy framework in place.
- Inputs and outputs—A structure requires inputs (typically information) before it can make informed decisions, and it produces outputs, e.g., decisions, other information, or requests for additional inputs.

ILLUSTRATIVE ORGANISATIONAL STRUCTURES IN COBIT 5

As mentioned in the discussion on the COBIT 5 process model, an illustrative COBIT 5 process reference model has been created and is described in detail in *COBIT 5: Enabling Processes*. The model includes RACI charts, which use a number of roles and structures. **Figure 33** describes these predefined roles and structures.

Notes:
- These do not necessarily have to correspond with actual functions that enterprises have implemented, but nonetheless they provide value in the sense that the described purpose of the structure or the role remains valid for most enterprises.
- The purpose of this table is not to prescribe a universal organisational chart for every enterprise. Rather, it should be seen as illustration.

Figure 33—COBIT 5 Roles and Organisational Structures	
Role/Structure	**Definition/Description**
Board	The group of the most senior executives and/or non-executive directors of the enterprise who are accountable for the governance of the enterprise and have overall control of its resources
CEO	The highest-ranking officer who is in charge of the total management of the enterprise
CFO	The most senior official of the enterprise who is accountable for all aspects of financial management, including financial risk and controls and reliable and accurate accounts
Chief Operating Officer (COO)	The most senior official of the enterprise who is accountable for the operation of the enterprise
CRO	The most senior official of the enterprise who is accountable for all aspects of risk management across the enterprise. An IT risk officer function may be established to oversee IT-related risk.
CIO	The most senior official of the enterprise who is responsible for aligning IT and business strategies and accountable for planning, resourcing and managing the delivery of IT services and solutions to support enterprise objectives
Chief Information Security Officer (CISO)	The most senior official of the enterprise who is accountable for the security of enterprise information in all its forms
Business Executive	A senior management individual accountable for the operation of a specific business unit or subsidiary
Business Process Owner	An individual accountable for the performance of a process in realising its objectives, driving process improvement and approving process changes
Strategy (IT Executive) Committee	A group of senior executives appointed by the board to ensure that the board is involved in, and kept informed of, major IT-related matters and decisions. The committee is accountable for managing the portfolios of IT-enabled investments, IT services and IT assets, ensuring that value is delivered and risk is managed. The committee is normally chaired by a board member, not by the CIO.
(Project and Programme) Steering Committees	A group of stakeholders and experts who are accountable for guidance of programmes and projects, including management and monitoring of plans, allocation of resources, delivery of benefits and value, and management of programme and project risk
Architecture Board	A group of stakeholders and experts who are accountable for guidance on enterprise architecture-related matters and decisions, and for setting architectural policies and standards
Enterprise Risk Committee	The group of executives of the enterprise who are accountable for the enterprise-level collaboration and consensus required to support enterprise risk management (ERM) activities and decisions. An IT risk council may be established to consider IT risk in more detail and advise the enterprise risk committee.
Head of HR	The most senior official of an enterprise who is accountable for planning and policies with respect to all human resources in that enterprise
Compliance	The function in the enterprise responsible for guidance on legal, regulatory and contractual compliance
Audit	The function in the enterprise responsible for provision of internal audits
Head of Architecture	A senior individual accountable for the enterprise architecture process
Head of Development	A senior individual accountable for IT-related solution development processes
Head of IT Operations	A senior individual accountable for the IT operational environments and infrastructure
Head of IT Administration	A senior individual accountable for IT-related records and responsible for supporting IT-related administrative matters
Programme and Project Management Office (PMO)	The function responsible for supporting programme and project managers, and gathering, assessing and reporting information about the conduct of their programmes and constituent projects
Value Management Office (VMO)	The function that acts as the secretariat for managing investment and service portfolios, including assessing and advising on investment opportunities and business cases, recommending value governance/management methods and controls, and reporting on progress on sustaining and creating value from investments and services
Service Manager	An individual who manages the development, implementation, evaluation and ongoing management of new and existing products and services for a specific customer (user) or group of customers (users)

Figure 33—COBIT 5 Roles and Organisational Structures *(cont.)*	
Role/Structure	**Definition/Description**
Information Security Manager	An individual who manages, designs, oversees and/or assesses an enterprise's information security
Business Continuity Manager	An individual who manages, designs, oversees and/or assesses an enterprise's business continuity capability, to ensure that the enterprise's critical functions continue to operate following disruptive events
Privacy Officer	An individual who is responsible for monitoring the risk and business impacts of privacy laws and for guiding and co-ordinating the implementation of policies and activities that will ensure that the privacy directives are met. Also called data protection officer.

Page intentionally left blank

COBIT 5 Enabler: Culture, Ethics and Behaviour

Culture, ethics and behaviour refers to the set of individual and collective behaviours within an enterprise. The specifics for the culture, ethics and behaviour enabler compared to the generic enabler description are shown in **figure 34**.

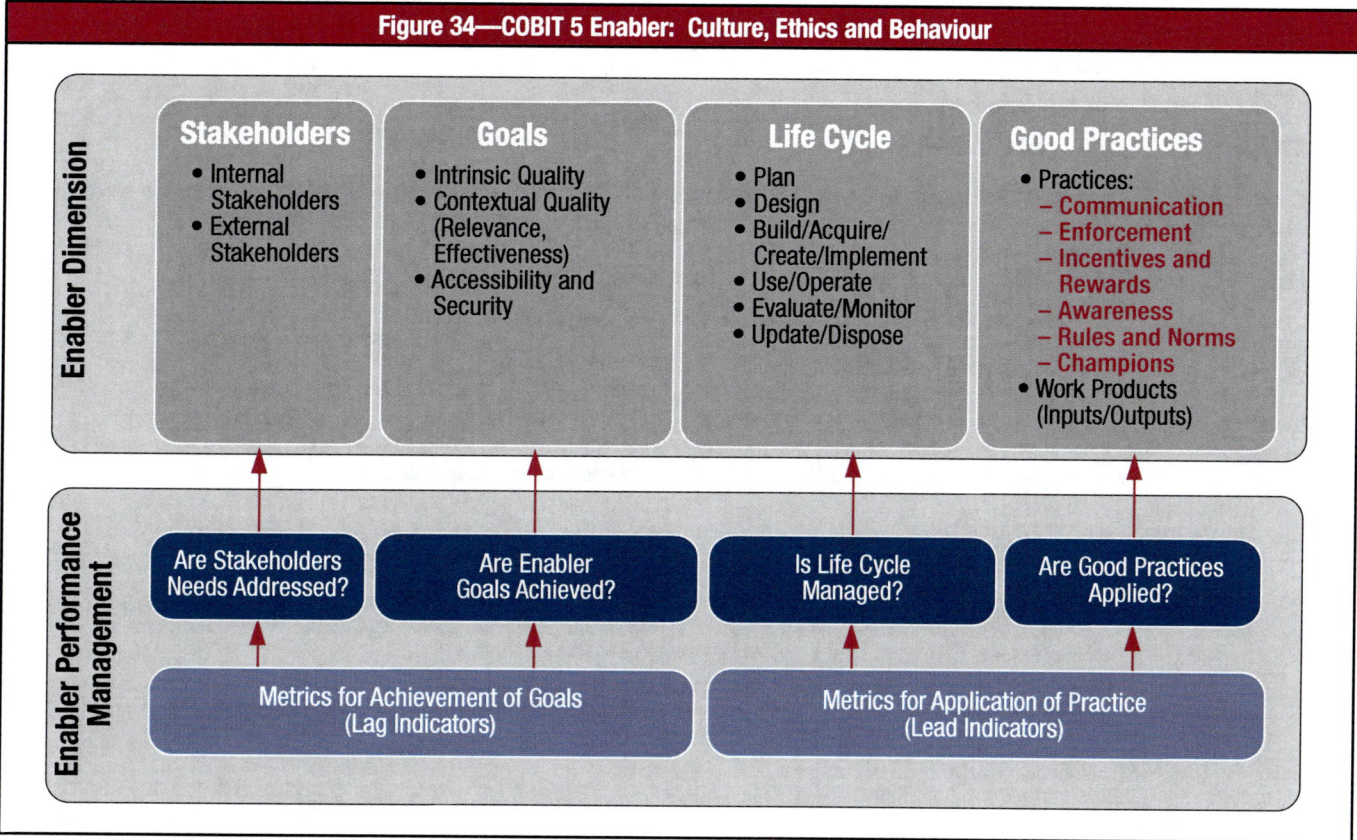

Figure 34—COBIT 5 Enabler: Culture, Ethics and Behaviour

The culture, ethics and behaviour model shows:
- **Stakeholders**—Culture, ethics and behaviour stakeholders can be internal and external to the enterprise. Internal stakeholders include the entire enterprise, external stakeholders include regulators, e.g., external auditors or supervisory bodies. Stakes are twofold: Some stakeholders, e.g., legal officers, risk managers, HR managers, remuneration boards and officers, deal with defining, implementing and enforcing desired behaviours, and others have to align with the defined rules and norms.
- **Goals**—Goals for the culture, ethics and behaviour enabler relate to:
  – Organisational ethics, determined by the values by which the enterprise wants to live
  – Individual ethics, determined by the personal values of each individual in the enterprise and depending to an important extent on external factors such as religion, ethnicity, socioeconomic background, geography and personal experiences
  – Individual behaviours, which collectively determine the culture of an enterprise. Many factors, such as the external factors mentioned above, but also interpersonal relationships in enterprises, personal objectives and ambitions, drive behaviours. Some types of behaviours that can be relevant in this context include:
    • Behaviour towards taking risk—How much risk does the enterprise feel it can absorb and which risk is it willing to take?
    • Behaviour towards following policy—To what extent will people embrace and/or comply with policy?
    • Behaviour towards negative outcomes—How does the enterprise deal with negative outcomes, i.e., loss events or missed opportunities? Will it learn from them and try to adjust, or will blame be assigned without treating the root cause?
- **Life cycle**—An organisational culture, ethical stance and individual behaviours, etc., all have their life cycles. Starting from an existing culture, an enterprise can identify required changes and work towards their implementation. Several tools—described in the good practices—can be used.
- **Good practices**—Good practices for creating, encouraging and maintaining desired behaviour throughout the enterprise include:
  – Communication throughout the enterprise of desired behaviours and the underlying corporate values
  – Awareness of desired behaviour, strengthened by the example behaviour exercised by senior management and other champions

- Incentives to encourage and deterrents to enforce desired behaviour. There is a clear link between individual behaviour and the HR reward scheme that an enterprise puts in place.
- Rules and norms, which provide more guidance on desired organisational behaviour. This links very clearly to the principles and policies that an enterprise puts in place.
- **Relationships with other enablers**—The links with other enablers include:
  - Processes can be designed to a level of perfection, but if the stakeholders of the process do not wish to execute the process activities as intended—i.e., if their behaviour is one of non-compliance—process outcomes will not be achieved.
  - Likewise, organisational structures can be designed and built according to the textbook, but if their decisions are not implemented—for reasons of different personal agendas, lack of incentives, etc.—they will not result in decent governance and management of enterprise IT.
  - Principles and policies are a very important communication mechanism for corporate values and the desired behaviour.

EXAMPLE 11—QUALITY IMPROVEMENT

An enterprise faces repeatedly serious quality problems with new applications. Despite the fact that a sound software project development methodology is in place, all too often software problems cause operational problems in day-to-day business.

An investigation shows that the development team members and management are evaluated and rewarded based on the timely delivery, within budget, of their projects. They are not measured against quality criteria or business benefits criteria. As a consequence, they focus diligently on delivery time and cost cutting during development, e.g., on testing time. The investigation also shows that compliance with the established methodology and procedures is virtually non-existent because it would take some more time from the development budget (in favour of quality). In addition, the organisational structure is such that the official involvement of development stops when the development has been handed over to the operations team. From then on, development's involvement is only indirect, through the established incident management and problem management processes.

The lesson learned is that better incentives must be used for solution development management and teams to encourage quality work.

EXAMPLE 12—IT-RELATED RISK

Some symptoms of an inadequate or problematic culture with regard to IT-related risk include:

- Misalignment between real risk appetite and translation into policies. Management's real values toward risk can be reasonably aggressive and risk-taking, whereas the policies that are created reflect a much more conservative attitude. Hence, there is a mismatch between values and the means to realise the values, inevitably leading to conflict. Conflicts may arise, for example, between the incentives set for management and the enforcement of misaligned policies.
- The existence of a 'blame culture'. This type of culture should by all means be avoided; it is the most effective inhibitor of relevant and efficient communication. In a blame culture, business units tend to point the finger at IT when projects are not delivered on time or do not meet expectations. In doing so, they fail to realise how the business unit's involvement up front affects project success. In extreme cases, the business unit may assign blame for a failure to meet the expectations that the unit never clearly communicated. The 'blame game' only detracts from effective communication across units, further fuelling delays. Executive leadership must identify and quickly control a blame culture if collaboration is to be fostered throughout the enterprise.

COBIT 5 Enabler: Information

Introduction—The Information Cycle

The information enabler deals with all information relevant for enterprises, not only automated information. Information can be structured or unstructured, formalised or informalised.

Information can be considered as being one stage in the 'information cycle' of an enterprise. In the information cycle (**figure 35**), business processes generate and process data, transforming them into information and knowledge, and ultimately generating value for the enterprise. The scope of the information enabler mainly concerns the 'information' phase in the information cycle, but the aspects of data and knowledge are also covered in COBIT 5.

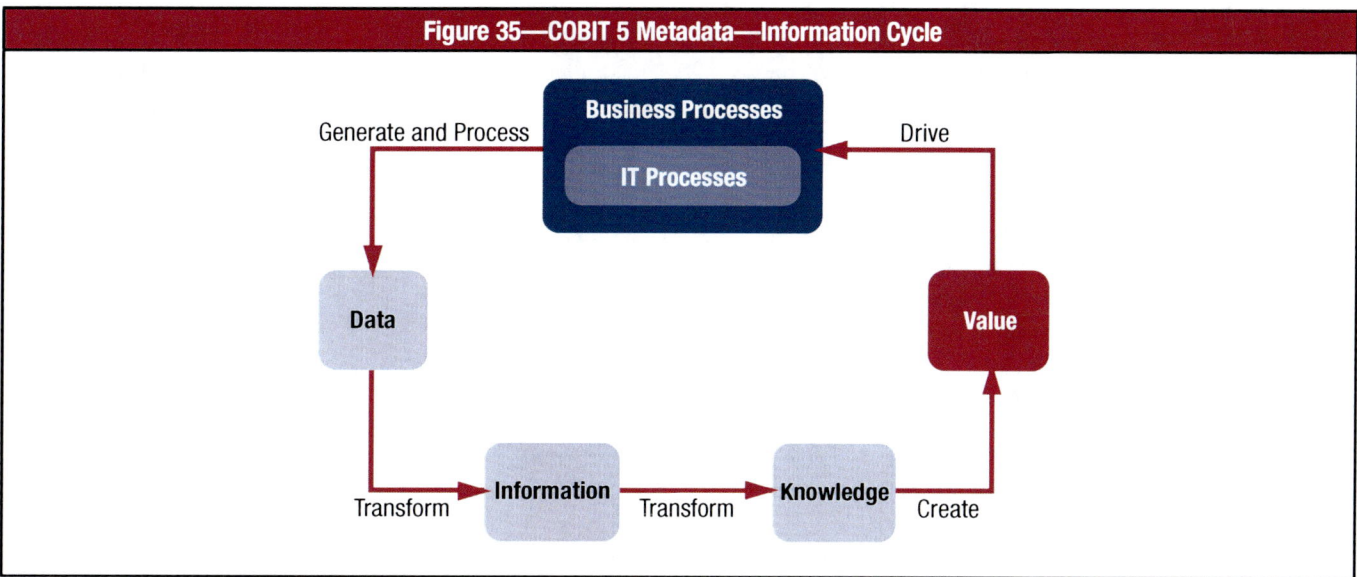

Figure 35—COBIT 5 Metadata—Information Cycle

COBIT 5 Information Enabler

The specifics for the information enabler compared to the generic enabler description are shown in **figure 36**.

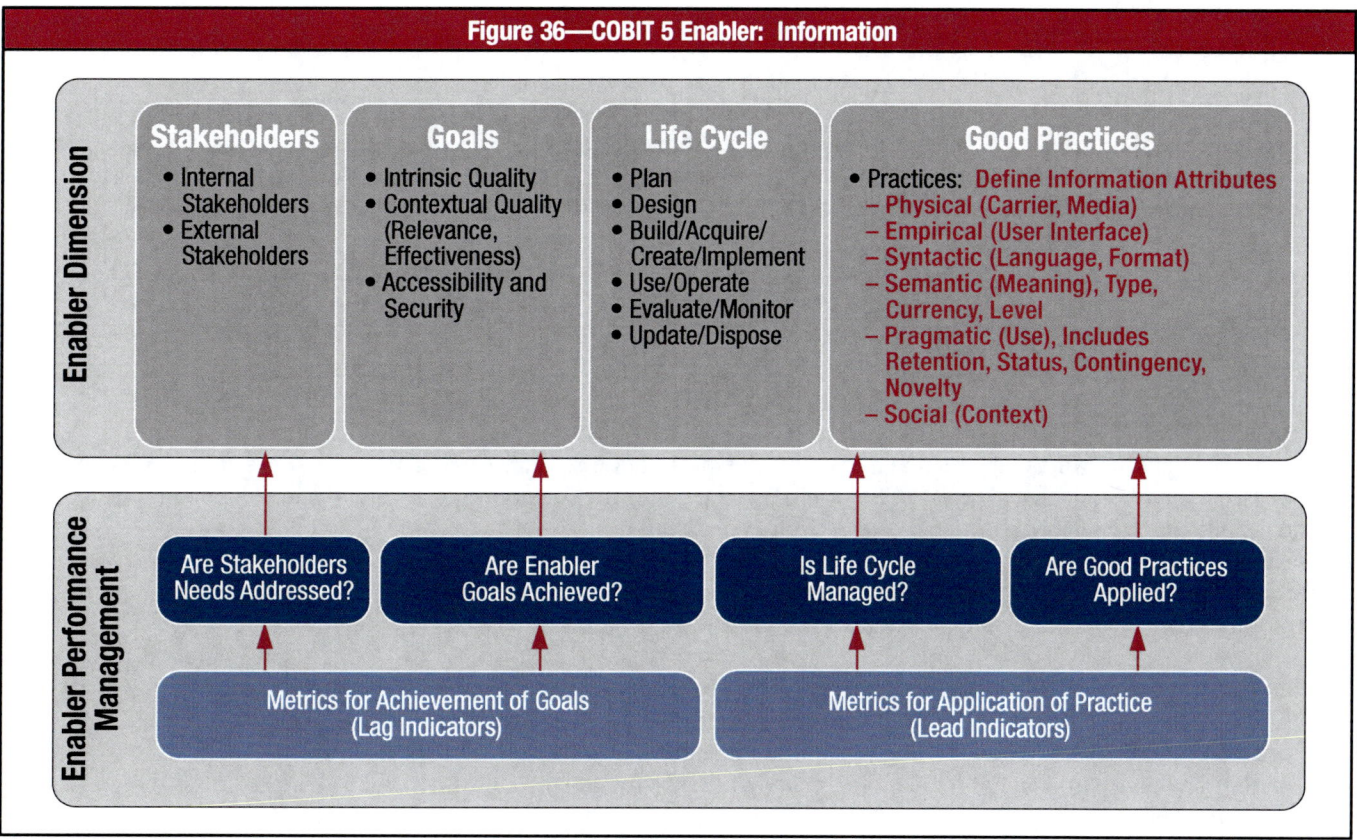

Figure 36—COBIT 5 Enabler: Information

The information model (IM) shows:
- **Stakeholders**—Can be internal or external to the enterprise. The generic model also suggests that, apart from identifying the stakeholders, their stakes need to be identified, i.e., why they care or are interested in the information.

With respect to which information stakeholders exist, different categories of roles in dealing with information are possible, ranging from detailed proposals—e.g., suggesting specific data or information roles such as architect, owner, steward, trustee, supplier, beneficiary, modeller, quality manager, security manager—to more general proposals—for instance, distinguishing amongst information producers, information custodians and information consumers:
  – Information producer, responsible for creating the information
  – Information custodian, responsible for storing and maintaining the information
  – Information consumer, responsible for using the information
  These categories refer to specific activities with regard to the information resource. Activities depend on the life cycle phase of the information; therefore, to find a category of roles that has an appropriate level of granularity for the IM, the information life cycle dimension of the IM can be used. This means that information stakeholder roles can be defined in terms of information life cycle phases, e.g., information planners, information obtainers, information users. At the same time, this means that the information stakeholder dimension is not an independent dimension; different life cycle phases have different stakeholders.

  Whereas the relevant roles depend on the information life cycle phase, the stakes can be related to information goals.
- **Goals**—The goals of information are divided into three subdimensions of quality:
  Intrinsic quality—The extent to which data values are in conformance with the actual or true values. It includes:
  – Accuracy—The extent to which information is correct and reliable
  – Objectivity—The extent to which information is unbiased, unprejudiced and impartial
  – Believability—The extent to which information is regarded as true and credible
  – Reputation—The extent to which information is highly regarded in terms of its source or content
  Contextual and representational quality—The extent to which information is applicable to the task of the information user and is presented in an intelligible and clear manner, recognising that information quality depends on the context of use. It includes:
  – Relevancy—The extent to which information is applicable and helpful for the task at hand
  – Completeness—The extent to which information is not missing and is of sufficient depth and breadth for the task at hand
  – Currency—The extent to which information is sufficiently up to date for the task at hand
  – Appropriate amount of information—The extent to which the volume of information is appropriate for the task at hand
  – Concise representation—The extent to which information is compactly represented
  – Consistent representation—The extent to which information is presented in the same format
  – Interpretability—The extent to which information is in appropriate languages, symbols and units, with clear definitions
  – Understandability—The extent to which information is easily comprehended
  – Ease of manipulation—The extent to which information is easy to manipulate and apply to different tasks
  Security/accessibility quality—The extent to which information is available or obtainable. It includes:
  – Availability/timeliness—The extent to which information is available when required, or easily and quickly retrievable
  – Restricted access—The extent to which access to information is restricted appropriately to authorised parties

  Appendix F provides a detailed description of how the COBIT 5 information quality criteria compare to the COBIT 4.1 information criteria. For example, integrity (as defined in COBIT 4.1) is covered by the information goals of completeness and accuracy.
- **Life cycle**—The full life cycle of information needs to be considered, and different approaches may be required for information in different phases of the life cycle. The COBIT 5 information enabler distinguishes the following phases:
  – **Plan**—The phase in which the creation and use of the information resource is prepared. Activities in this phase may refer to the identification of objectives, the planning of the information architecture, and the development of standards and definitions, e.g., data definitions, data collection procedures.
  – **Design**
  – **Build/acquire**—The phase in which the information resource is acquired. Activities in this phase may refer to the creation of data records, the purchase of data and the loading of external files.
  – **Use/operate**, which includes:
    • Store—The phase in which information is held electronically or in hard copy (or even just in human memory). Activities in this phase may refer to the storage of information in electronic form (e.g., electronic files, databases, data warehouses) or as hard copy (e.g., paper documents).
    • Share—The phase in which information is made available for use through a distribution method. Activities in this phase may refer to the processes involved in getting the information to places where it can be accessed and used, e.g., distributing documents by email. For electronically held information, this life cycle phase may largely overlap with the store phase, e.g., sharing information through database access, file/document servers.

• Use—The phase in which information is used to accomplish goals. Activities in this phase may refer to all kinds of information usage (e.g., managerial decision making, running automated processes), and may also include activities such as information retrieval and converting information from one form to another.

According to the Taking Governance Forward view, information is an enabler for enterprise governance; hence, information use as defined in the IM can be thought of as the purposes for which enterprise governance stakeholders need information when assuming their roles, fulfilling their activities and interacting with each other.

These roles, activities and relationships are captured in **figure 8**. The interactions between stakeholders require information flows whose purposes are indicated in the schema: accountability, delegation, monitoring, direction setting, alignment, execution and control.
  – Monitor—The phase in which it is ensured that the information resource continues to work properly, i.e., to be valuable. Activities in this phase may refer to keeping information up to date as well as other kinds of information management activities, e.g., enhancing, cleansing, merging, removing duplicate information data in data warehouses.
  – Dispose—The phase in which the information resource is discarded when it is no longer of use. Activities in this phase may refer to information archiving or destroying.
• **Best practice**—The concept of information is understood differently in different disciplines such as economics, communication theory, information science, knowledge management and information systems; therefore, there is no universally agreed-on definition regarding what information is. The nature of information can, however, be clarified through defining and describing its properties.

The following scheme is proposed to structure information's different properties: it consists of six levels or layers to define and describe properties of information. These six levels present a continuum of attributes, ranging from the physical world of information, where attributes are linked to information technologies and media for information capturing, storing, processing, distribution and presentation, to the social world of information use, comprehension and action.

The following descriptions can be given to the layers and information attributes:
• **Physical world layer**—The world where all phenomena that can be empirically observed take place
  – Information carrier/media—The attribute that identifies the physical carrier of the information, e.g., paper, electric signals, sound waves
• **Empiric layer**—The empirical observation of the signs used to encode information and their distinction from each other and from background noise
  – Information access channel—The attribute that identifies the access channel of the information, e.g., user interfaces
• **Syntactic layer**—The rules and principles for constructing sentences in natural or artificial languages. Syntax refers to the form of information.
  – Code/language—Attribute that identifies the representational language/format used for encoding the information and the rules for combining the symbols of the language to form syntactic structures.
• **Semantic layer**—The rules and principles for constructing meaning out of syntactic structures. Semantics refers to the meaning of information.
  – Information type—The attribute that identifies the kind of information, e.g., financial vs. non-financial information, internal vs. external origin of the information, forecasted/predicted vs. observed values, planned vs. realised values
  – Information currency—The attribute that identifies the time horizon referred to by the information, i.e., information on the past, the present or the future
  – Information level—The attribute that identifies the degree of detail of the information, e.g., sales per year, quarter, month
• **Pragmatic layer**—The rules and structures for constructing larger language structures that fulfil specific purposes in human communication. Pragmatics refers to the use of information.
  – Retention period—The attribute that identifies how long information can be retained before it is destroyed
  – Information status—The attribute that identifies whether the information is operational or historical
  – Novelty—The attribute that identifies whether the information creates new knowledge or confirms existing knowledge, i.e., information vs. confirmation
  – Contingency—The attribute that identifies the information that is required to precede this information (for it to be considered as information)
• **Social world layer**—The world that is socially constructed through the use of language structures at the pragmatic level of semiotics, e.g., contracts, law, culture
  – Context—The attribute that identifies the context in which the information makes sense, is used, has value, etc., e.g., cultural context, subject domain context

Further considerations about information—Investments in information and related technology are based on business cases, which include cost-benefit analysis. Costs and benefits refer not only to tangible, measurable factors, but they also take into account intangible factors such as competitive advantage, customer satisfaction and technology uncertainty. It is only when the information resource is applied or used that an enterprise generates benefits from it, so the value of information is determined solely through its use (internally or by selling it), and information has no intrinsic value. It is only through putting information into action that value can be generated.

The IM is a new model and is very rich in terms of different components. It will be further developed in a separate publication. To make it more tangible for the COBIT 5 user, and to make its relevance more clear in the context of the overall COBIT 5 framework, examples 13, 14 and 15 of possible use of the IM are provided.

EXAMPLE 13—INFORMATION MODEL USED FOR INFORMATION SPECIFICATIONS

When developing a new application, the IM can be used to assist with the specifications of the application and the associated information or data models.

The information attributes of the IM can be used to define specifications for the application and the business processes that will use the information.

For example, the design and specifications of the new system need to specify:
- **Physical layer**—Where will information be stored?
- **Empirical layer**—How can the information be accessed?
- **Syntactical layer**—How will the information be structured and coded?
- **Semantic layer**—What sort of information is it? What is the information level?
- **Pragmatic layer**—What are the retention requirements? What other information is required for this information to be useful and usable?

Looking at the stakeholder dimension combined with the information life cycle, one can define who will need what type of access to the data during which phase of the information life cycle.

When the application is tested, testers can look at the information quality criteria to develop a comprehensive set of test cases.

EXAMPLE 14—INFORMATION MODEL USED TO DETERMINE REQUIRED PROTECTION

Security groups within the enterprise can benefit from the attributes dimension of the IM. Indeed, when charged with protection of information, they need to look at:
- **Physical layer**—How and where is information physically stored?
- **Empirical layer**—What are the access channels to the information?
- **Semantic layer**—What type of information is it? Is the information current or relating to the past or to the future?
- **Pragmatic layer**—What are the retention requirements? Is information historic or operational?

Using these attributes will allow the user to determine the level of protection and the protection mechanisms required.

Looking at another dimension of the IM, security professionals can also consider the information life cycle stages, because information needs to be protected during all phases of the life cycle. Indeed, security starts at the information planning phase, and implies different protection mechanisms for storing, sharing and disposition of information. The IM ensures that information is protected during the full life cycle of the information.

EXAMPLE 15—INFORMATION MODEL USED TO DETERMINE EASE OF DATA USE

When performing a review of a business process (or an application), the IM can be used to assist with a general review of the information processed and delivered by the process, and of the underlying information systems. The quality criteria can be used to assess the extent to which information is available—whether the information is complete, available on a timely basis, factually correct, relevant, available in the appropriate amount. One can also consider the accessibility criteria—whether the information is accessible when required and adequately protected.

The review can be even further extended to include representation criteria, e.g., the ease with which the information can be understood, interpreted, used and manipulated.

A review that uses the information quality criteria of the IM provides an enterprise with a comprehensive and complete view on the current information quality within a business process.

COBIT 5 Enabler: Services, Infrastructure and Applications

Service capabilities refer to resources such as applications and infrastructures that are leveraged in the delivery of IT-related services.

The specifics for the service capabilities enabler compared to the generic enabler description are shown in **figure 37**.

The services, infrastructure and applications model shows:
- **Stakeholders**—Service capabilities (the combined term for services, infrastructure and applications) stakeholders can be internal and external. Services can be delivered by internal or external parties—internal IT departments, operations managers, outsourcing providers. Users of services can also be internal— business users—and external to the enterprise—partners, clients, suppliers. The stakes of each of the stakeholders need to be identified and will either be focussed on delivering adequate services or on receiving requested services from providers.
- **Goals**—Goals of the service level capability will be expressed in terms of services—applications, infrastructure, technology—and service levels, considering which services and service levels are most economical for the enterprise. Again, goals will relate to the services and how they are provided, as well as their outcomes, i.e., contribution towards successfully supported business processes.
- **Life cycle**—Service capabilities have a life cycle. The future or planned service capabilities are typically described in a target architecture. It covers the building blocks, such as future applications and the target infrastructure model, and also describes the linkages and relationships amongst these building blocks.

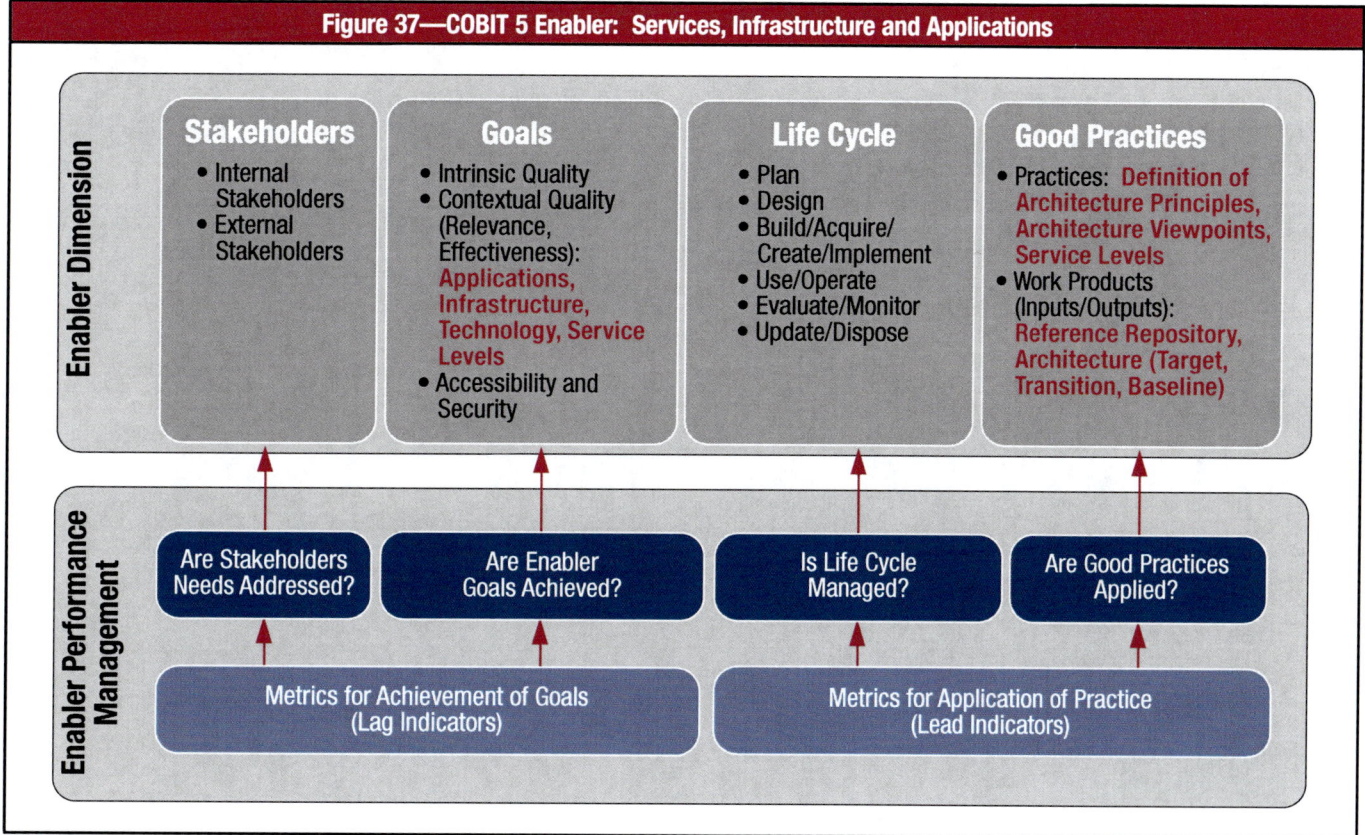

Figure 37—COBIT 5 Enabler: Services, Infrastructure and Applications

The current service capabilities that are used or operated to deliver current IT services are described in a baseline architecture. Depending on the time frame of the target architecture, a transition architecture may also be defined, which shows the enterprise at incremental states between the target and baseline architectures.
- **Good practices**—Good practice for service capabilities includes:
  - Definition of architecture principles—Architecture principles are overall guidelines that govern the implementation and use of IT-related resources within the enterprise. Examples of potential architecture principles are:
    - **Reuse**—Common components of the architecture should be used when designing and implementing solutions as part of the target or transition architectures.
    - **Buy vs. build**—Solutions should be purchased unless there is an approved rationale for developing them internally.

- **Simplicity**—The enterprise architecture should be designed and maintained to be as simple as possible while still meeting enterprise requirements.
  - **Agility**—The enterprise architecture should incorporate agility to meet changing business needs in an effective and efficient manner.
  - **Openness**—The enterprise architecture should leverage open industry standards.
- The enterprise's definition of the most appropriate architecture viewpoints to meet the needs of different stakeholders. These are the models, catalogues and matrices used to describe the baseline, target or transition architectures; for example, an application architecture could be described through an application interface diagram, which shows the applications in use (or planned) and the interfaces amongst them.
- Having an architecture repository, which can be used to store different types of architectural outputs, including architecture principles and standards, architecture reference models, and other architecture deliverables, and which defines the building blocks of services such as:
  - Applications, providing business functionality
  - Technology infrastructure, including hardware, system software and networking infrastructure
  - Physical infrastructure
- Service levels that need to be defined and achieved by the service providers

External good practice for architecture frameworks and service capabilities exist. These are guidelines, templates or standards that could be used to fast-track the creation of architecture deliverables. Examples are:
- TOGAF[16] provides a Technical Reference Model and an Integrated Information Infrastructure Reference Model.
- ITIL provides comprehensive guidance on how to design and operate services.
- **Relationships with other enablers**—The links with other enablers include:
  - Information is one of the service capabilities, and service capabilities are leveraged through processes to deliver internal and external services.
  - Cultural and behavioural aspects are also relevant when a service-oriented culture needs to be built.
  - Within COBIT 5, the inputs and outputs of the management practices and activities could include service capabilities, which are required as inputs or delivered as outputs.

---

[16] *www.opengroup.org/togaf*

COBIT 5 Enabler: People, Skills and Competencies

The specifics for the people, skills and competencies enabler compared to the generic enabler description are shown in **figure 38**.

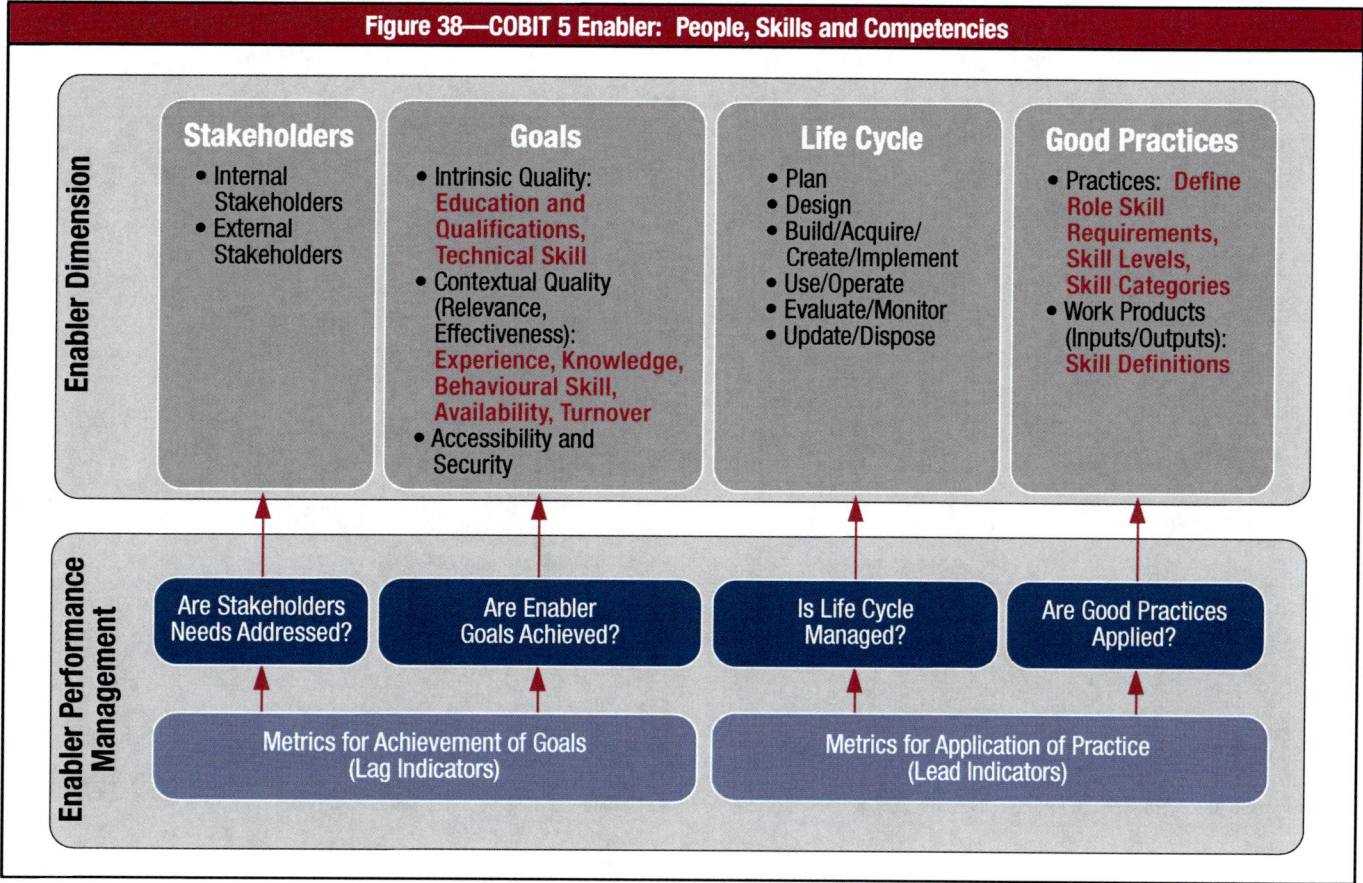

Figure 38—COBIT 5 Enabler: People, Skills and Competencies

The people, skills and competencies model shows:
- **Stakeholders**—Skills and competencies stakeholders are internal and external to the enterprise. Different stakeholders assume different roles—business managers, project managers, partners, competitors, recruiters, trainers, developers, technical IT specialists, etc.—and each role requires a distinct skill set.
- **Goals**—Goals for skills and competencies relate to education and qualification levels, technical skills, experience levels, knowledge and behavioural skills required to provide and perform successfully process activities, organisational roles, etc. Goals for people include correct levels of staff availability and turnover rate.
- **Life cycle:**
  – Skills and competencies have a life cycle. An enterprise has to know what its current skill base is, and plan what it needs to be. This is influenced by (amongst other issues) the strategy and goals of the enterprise. Skills need to be developed (e.g., through training) or acquired (e.g., through recruitment) and deployed in the various roles within the organisational structure. Skills may need to be disposed of, e.g., if an activity is automated or outsourced.
  – Periodically, such as on an annual basis, the enterprise needs to assess the skill base to understand the evolution that has occurred, which will feed into the planning process for the next period.
  – This assessment can also feed into the reward and recognition process for human resources.
- **Good practices:**
  – Good practice for skills and competencies includes defining the need for objective skill requirements for each role played by the various stakeholders. This can be described through different skill levels in different skill categories. For each appropriate skill level in each skill category, a skill definition should be available. The skill categories correspond with the IT-related activities undertaken, e.g., information management, business analysis.

– Other good practice:
  • There are external sources of good practice, such as the Skills Framework for the Information Age (SFIA),[17] which provides comprehensive skill definitions.
  • Examples of potential skill categories, mapped to COBIT 5 process domains, are shown in **figure 39**.

Figure 39—COBIT 5 Skill Categories	
Process Domain	**Examples of Skill Categories**
Evaluate, Direct and Monitor (EDM)	• Governance of enterprise IT
Align, Plan and Organise (APO)	• IT policy formulation • IT strategy • Enterprise architecture • Innovation • Financial management • Portfolio management
Build, Acquire and Implement (BAI)	• Business analysis • Project management • Usability evaluation • Requirements definition and management • Programming • System ergonomics • Software decommissioning • Capacity management
Deliver, Service and Support (DSS)	• Availability management • Problem management • Service desk and incident management • Security administration • IT operations • Database administration
Monitor, Evaluate and Assess (MEA)	• Compliance review • Performance monitoring • Controls audit

• **Relationships with other enablers**—The links with other enablers include:
  – Skills and competence are required to perform process activities and take decisions in organisational structures. Conversely, some processes are aimed at supporting the life cycle of skills and competencies.
  – There is also a link to culture, ethics and behaviour through behavioural skills, which drive individual behaviour and are influenced by individual ethics and organisational ethics.
  – Skills definitions are also information, for which best practices of the information enabler need to be considered.

APPENDIX H
GLOSSARY

TERM	DEFINITION
Accountable party (RACI)	The individual, group or entity that is ultimately responsible for a subject matter, process or scope In a RACI chart, answers the question: **Who accounts for the success of the task?**
Accountability of governance	Governance ensures that enterprise objectives are achieved by evaluating stakeholder needs, conditions and options; setting direction through prioritisation and decision making; and monitoring performance, compliance and progress against plans. In most enterprises, governance is the responsibility of the board of directors, under the leadership of the chairperson.
Activity	In COBIT, the main action taken to operate the process. Guidance to achieve management practices for successful governance and management of enterprise IT. Activities: • Describe a set of necessary and sufficient action-oriented implementation steps to achieve a Governance Practice or Management Practice • Consider the inputs and outputs of the process • Are based on generally accepted standards and good practices • Support establishment of clear roles and responsibilities • Are non-prescriptive and need to be adapted and developed into specific procedures appropriate for the enterprise
Alignment	A state where the enablers of governance and management of enterprise IT support the goals and strategies of the enterprise
Application architecture	Description of the logical grouping of capabilities that manage the objects necessary to process information and support the enterprise's objectives
Architecture board	A group of stakeholders and experts who are accountable for guidance on enterprise-architecture-related matters and decisions, and for setting architectural policies and standards
Authentication	The act of verifying the identity of a user and the user's eligibility to access computerised information Scope Note: Assurance: Authentication is designed to protect against fraudulent logon activity. It can also refer to the verification of the correctness of a piece of data.
Baseline architecture	The existing description of the fundamental underlying design of the components of the business system before entering a cycle of architecture review and redesign
Benefits realisation	One of the objectives of governance. The bringing about of new benefits for the enterprise, the maintenance and extension of existing forms of benefits, and the elimination of those initiatives and assets that are not creating sufficient value.
Business continuity	Preventing, mitigating and recovering from disruption. The terms 'business resumption planning', 'disaster recovery planning' and 'contingency planning' also may be used in this context; they focus on recovery aspects of continuity, and for that reason the 'resilience' aspect should also be taken into account.
Business goal	The translation of the enterprise's mission from a statement of intention into performance targets and results
Business process control	The policies, procedures, practices and organisational structures designed to provide reasonable assurance that a business process will achieve its objectives
Chargeback	The redistribution of expenditures to the units within a company that gave rise to them Scope Note: Chargeback is important because without such a policy, misleading views may be given as to the real profitability of a product or service, as certain key expenditures will be ignored or calculated according to an arbitrary formula.

TERM	DEFINITION
COBIT	1. COBIT 5: Formerly known as Control Objectives for Information and related Technology (COBIT); now used only as the acronym in its fifth iteration. A complete, internationally accepted framework for governing and managing enterprise information and technology (IT) that supports enterprise executives and management in their definition and achievement of business goals and related IT goals. COBIT describes five principles and seven enablers that support enterprises in the development, implementation, and continuous improvement and monitoring of good IT-related governance and management practices. Scope Note: Earlier versions of COBIT focused on control objectives related to IT processes, management and control of IT processes and IT governance aspects. Adoption and use of the COBIT framework are supported by guidance from a growing family of supporting products. (See *www.isaca.org/cobit* for more information.) 2. COBIT 4.1 and earlier: Formerly known as Control Objectives for Information and related Technology (COBIT). A complete, internationally accepted process framework for IT that supports business and IT executives and management in their definition and achievement of business goals and related IT goals by providing a comprehensive IT governance, management, control and assurance model. COBIT describes IT processes and associated control objectives, management guidelines (activities, accountabilities, responsibilities and performance metrics) and maturity models. COBIT supports enterprise management in the development, implementation, continuous improvement and monitoring of good IT-related practices. Scope Note: Adoption and use of the COBIT framework are supported by guidance for executives and management (*Board Briefing on IT Governance, 2nd Edition*), IT governance implementers (*COBIT Quickstart, 2nd Edition; IT Governance Implementation Guide: Using COBIT and Val IT, 2nd Edition;* and *COBIT Control Practices: Guidance to Achieve Control Objectives for Successful IT Governance*), and IT assurance and audit professionals (*IT Assurance Guide Using COBIT*). Guidance also exists to support its applicability for certain legislative and regulatory requirements (e.g., *IT Control Objectives for Sarbanes-Oxley, IT Control Objectives for Basel II*) and its relevance to information security (*COBIT Security Baseline*). COBIT is mapped to other frameworks and standards to illustrate complete coverage of the IT management life cycle and support its use in enterprises using multiple IT-related framework and standards.
Code of ethics	A document designed to influence individual and organisational behaviour of employees by defining organisational values and the rules to be applied in certain situations. It is adopted to assist those in the enterprise called upon to make decisions understand the difference between 'right' and 'wrong' and to apply this understanding to their decisions.
Competence	The ability to perform a specific task, action or function successfully
Consulted party (RACI)	Refers to those people whose opinions are sought on an activity (two-way communication) In a RACI chart, answers the question: **Who is providing input?** Key roles that provide input. Note that it is up to the accountable and responsible roles to obtain information from other units or external partners, too; however, inputs from the roles listed are to be considered and, if required, appropriate action has to be taken for escalation, including the information of the process owner and/or the steering committee.
Context	The overall set of internal and external factors that might influence or determine how an enterprise, entity, process or individual acts Scope Note: Context includes: • Technology context—Technological factors that affect an organisation's ability to extract value from data • Data context—Data accuracy, availability, currency and quality • Skills and knowledge—General experience, and analytical, technical and business skills • Organisational and cultural context—Political factors, and whether the organisation prefers data to intuition • Strategic context—Strategic objectives of the enterprise

TERM	DEFINITION
Control	The means of managing risk, including policies, procedures, guidelines, practices or organisational structures, which can be of an administrative, technical, management or legal nature. Also used as a synonym for safeguard or countermeasure.
Culture	A pattern of behaviours, beliefs, assumptions, attitudes and ways of doing things
Driver	External and internal factors that initiate and affect how an enterprise or individuals act or change
Enterprise goal	See Business goal
Enterprise governance	A set of responsibilities and practices exercised by the board and executive management with the goal of providing strategic direction, ensuring that objectives are achieved, ascertaining that risk is managed appropriately and verifying that the enterprise's resources are used responsibly. It could also mean a governance view focussing on the overall enterprise; the highest-level view of governance to which all others must align.
Full economic life cycle	The period of time during which material business benefits are expected to arise from, and/or during which material expenditures (including investments, running and retirement costs) are expected to be incurred by, an investment programme
Good practice	A proven activity or process that has been successfully used by multiple enterprises and has been shown to produce reliable results
Governance	Governance ensures that stakeholder needs, conditions and options are evaluated to determine balanced, agreed-on enterprise objectives to be achieved; setting direction through prioritisation and decision making; and monitoring performance and compliance against agreed-on direction and objectives.
Governance/management practice	For each COBIT process, the governance and management practices provide a complete set of high-level requirements for effective and practical governance and management of enterprise IT. They are statements of actions from governance bodies and management.
Governance enabler	Something (tangible or intangible) that assists in the realisation of effective governance
Governance framework	A framework is a basic conceptual structure used to solve or address complex issues; an enabler of governance; a set of concepts, assumptions and practices that define how something can be approached or understood, the relationships amongst the entities involved, the roles of those involved, and the boundaries (what is and is not included in the governance system). Examples: COBIT and COSO's *Internal Control—Integrated Framework*
Governance of enterprise IT	A governance view that ensures that information and related technology support and enable the enterprise strategy and the achievement of enterprise objectives. It also includes the functional governance of IT, i.e., ensuring that IT capabilities are provided efficiently and effectively.
Information	An asset that, like other important business assets, is essential to an enterprise's business. It can exist in many forms: printed or written on paper, stored electronically, transmitted by post or electronically, shown on films, or spoken in conversation.
Informed party (RACI)	Refers to those people who are kept up to date on the progress of an activity (one-way communication) In a RACI chart, answers the question: **Who is receiving information?** Roles who are informed of the achievements and/or deliverables of the task. The role in 'accountable', of course, should always receive appropriate information to oversee the task, as do the responsible roles for their area of interest.
Inputs and outputs	The process work products/artefacts considered necessary to support operation of the process. They enable key decisions, provide a record and audit trail of process activities, and enable follow-up in the event of an incident. They are defined at the key management practice level, may include some work products used only within the process and are often essential inputs to other processes. The illustrative COBIT 5 inputs and outputs should not be regarded as an exhaustive list since additional information flows could be defined depending on a particular enterprise's environment and process framework.
Investment portfolio	The collection of investments being considered and/or being made

TERM	DEFINITION
IT application	Electronic functionality that constitutes parts of business processes undertaken by, or with the assistance of, IT
IT goal	A statement describing a desired outcome of enterprise IT in support of enterprise goals. An outcome can be an artefact, a significant change of a state or a significant capability improvement.
IT service	The day-to-day provision to customers of IT infrastructure and applications and support for their use. Examples include service desk, equipment supply and moves, and security authorisations.
Management	Management plans, builds, runs and monitors activities in alignment with the direction set by the governance body to achieve the enterprise objectives.
Model	A way to describe a given set of components and how those components relate to each other to describe the main workings of an object, system, or concept
Metric	A quantifiable entity that allows the measurement of the achievement of a process goal. Metrics should be SMART—specific, measurable, actionable, relevant and timely. Complete metric guidance defines the unit used, measurement frequency, ideal target value (if appropriate) and also the procedure to carry out the measurement and the procedure for the interpretation of the assessment.
Objective	Statement of a desired outcome
Organisational structure	An enabler of governance and of management. Includes the enterprise and its structures, hierarchies and dependencies. Example: Steering committee
Output	See Inputs and outputs
Owner	Individual or group that holds or possesses the rights of and the responsibilities for an enterprise, entity or asset, e.g., process owner, system owner
Policy	Overall intention and direction as formally expressed by management
Principle	An enabler of governance and of management. Comprises the values and fundamental assumptions held by the enterprise, the beliefs that guide and put boundaries around the enterprise's decision making, communication within and outside the enterprise, and stewardship— caring for assets owned by another. Example: Ethics charter, social responsibility charter
Process	Generally, a collection of practices influenced by the enterprise's policies and procedures that takes inputs from a number of sources (including other processes), manipulates the inputs and produces outputs (e.g., products, services) Scope note: Processes have clear business reasons for existing, accountable owners, clear roles and responsibilities around the execution of the process, and the means to measure performance.
Process (capability) attribute	ISO/IEC 15504: A measurable characteristic of process capability applicable to any process
Process capability	ISO/IEC 15504: A characterization of the ability of a process to meet current or projected business goals
Process goal	A statement describing the desired outcome of a process. An outcome can be an artefact, a significant change of a state or a significant capability improvement of other processes.
Programme and project management office (PMO)	The function responsible for supporting programme and project managers, and gathering, assessing and reporting information about the conduct of their programmes and constituent projects
Quality	Being fit for purpose (achieving intended value)
RACI chart	Illustrates who is responsible, accountable, consulted and informed within an organisational framework
Resource	Any enterprise asset that can help the organisation achieve its objectives

TERM	DEFINITION
Resource optimisation	One of the governance objectives. Involves effective, efficient and responsible use of all resources—human, financial, equipment, facilities, etc.
Responsible party (RACI)	Refers to the person who must ensure that activities are completed successfully In a RACI chart, answers the question: **Who is getting the task done?** Roles taking the main operational stake in fulfilling the activity listed and creating the intended outcome
Risk	The combination of the probability of an event and its consequence (ISO/IEC 73)
Risk management	One of the governance objectives. Entails recognising risk; assessing the impact and likelihood of that risk; and developing strategies, such as avoiding the risk, reducing the negative effect of the risk and/or transferring the risk, to manage it within the context of the enterprise's risk appetite.
Service catalogue	Structured information on all IT services available to customers
Services	See IT service
Skill	The learned capacity to achieve predetermined results
Stakeholder	Anyone who has a responsibility for, an expectation from or some other interest in the enterprise—e.g., shareholders, users, government, suppliers, customers and the public
System of internal control	The policies, standards, plans and procedures, and organisational structures designed to provide reasonable assurance that enterprise objectives will be achieved and undesired events will be prevented or detected and corrected
Value creation	The main governance objective of an enterprise, achieved when the three underlying objectives (benefits realisation, risk optimisation and resource optimisation) are all balanced

Page intentionally left blank

Page intentionally left blank